Student Self-Study Guide

Fundamentals of Chemistry

Karl Kumli
California State University, Chico

D. Van Nostrand Company
New York Cincinnati Toronto London Melbourne

D. Van Nostrand Company Regional Offices:
New York   Cincinnati

D. Van Nostrand Company International Offices:
London   Toronto   Melbourne

ISBN: 0-442-27349-5

Published by D. Van Nostrand Company
135 West 50th Street, New York, NY 10020

10 9 8 7 6 5 4 3 2 1

Preface

The purpose of this study guide is to serve as a silent tutor that will help you master some of the basic principles of chemistry. It accompanies <u>Fundamentals of Chemistry</u>, by Karl Kumli. The sequence of chapters and topics in this guide is identical to that of the textbook.

Use this guide as you would use any tutor. A tutor can help you learn a subject only if you help yourself. First, there is no substitute for a lecturer or instructor. The instructor represents your major contact with the subject, directing your study through lectures and discussions and evaluating your progress in the course. The primary requirement in any course is that you attend <u>all</u> of the lectures, discussions, and laboratories. To prepare for the lectures, read over the material to be discussed, note those points that seem difficult, and formulate questions about what you do not understand. After the lecture, reread those sections of the text that have been covered in lecture and <u>work</u> <u>problems</u>. Since this course is meant to help you improve your ability to handle the quantitative aspects of chemistry, the more problems you work, the better.

This study guide is designed to help you through the more difficult parts of the textbook. Each chapter of the study guide contains the following: a list of

learning objectives for each chapter of the text; a
list of important terms and concepts with references
to the section of the text in which they are discussed;
a set of problems similar in nature to those discussed
in the text; detailed solutions for these problems;
short self-tests over each chapter with answers for
the test.

Contents

CHAPTER 1   Introduction   1

CHAPTER 2   Math Review and Scientific Measurements 3

CHAPTER 3   Matter and Energy   22

CHAPTER 4   Atomic Structure   31

CHAPTER 5   The Periodic Table   45

CHAPTER 6   Chemical Bonds Between Atoms   58

CHAPTER 7   Stoichiometry, The Quantitative Aspects of Chemical Reactions   74

CHAPTER 8   The Behavior of Gases   93

CHAPTER 9   Liquids and Solids   110

CHAPTER 10   Solutions   123

CHAPTER 11   Aqueous Solutions of Acids, Bases, and Salts   139

CHAPTER 12   Rates of Reactions and Chemical Equilibrium   161

CHAPTER 13   Oxidation-Reduction Reactions   173

1 Introduction

As the title states, this chapter is simply a narrative introduction to science in general and chemistry specifically.

OBJECTIVES
1. From this chapter you should begin to develop a vocabulary of chemical terms and learn precise meanings for each of these terms. (Sections 1.1, 1.2, and 1.3)
2. You should develop an initial plan of study for this course. (Section 1.6)

IMPORTANT
TERMS
AND
CONCEPTS

Section 1.1

Science   the knowledge of facts, phenomena, laws, and proximate causes of a subject gained and verified by exact observations, organized experiments, and critical thinking.

Section 1.2

Scientific method   the method of the acquisition of knowledge that involves observation, correlation, prediction, and verification.

Inductive reasoning   the process of development of assumptions from a collection of facts.

Hypothesis   assumptions made from observed facts.

1

Theory   assumptions explaining observations which have
    been verified by such a large number of facts that
    they are almost irrefutable.

SECTION 1.3

Chemistry   the science that studies matter, the trans-
    formation of matter and the energy changes accom-
    panying these transformations.

Matter   anything that has mass and occupies space.

QUESTIONS
AND
PROBLEMS

SECTION 1.2

1. Relate the discovery of atmospheric pressure des-
   cribed in this section to the scientific method.

Answer:

1. Observation of Facts
     Water rises only 34 feet in a vacuum.
     Mercury rises only 30 inches in a vacuum.
     Air has mass.
     The height of mercury in a Torricelli tube varies
       from day to day.

   Hypothesis
     The weight of the air of the atmosphere pushes the
       water and the mercury into the vacuum.

   Verification
     The height of mercury in a Torricelli tube de-
       creases as the altitude increases or as the
       weight of atmospheric air decreases.

2 Math Review & Scientific Measurements

In this chapter, you should review all of the simple mathematical operations, the use of negative numbers, the use of exponential numbers, the use of scientific notation, the use of logarithms, the solving of simple equations, and the use and interpretation of graphs. You should learn the International System of Measurements and the Metric System of Measurements.

OBJECTIVES
1. After completing the math review, you should be able to work all of the problems of addition, subtraction, multiplication, and division using a pocket calculator (Section 2.1).
2. You should be able to perform all of the arithmetical manipulations which involve negative numbers (Section 2.2).
3. You should be able to solve simple first order equations for a given unknown (Section 2.3).
4. You should be able to work with all exponential numbers and perform the arithmetical operations using exponential numbers (Section 2.4).
5. You should be able to perform the following operations with scientific notation of numbers: (It is okay if you use a scientific pocket calculator.)
   a. Convert decimal numbers to scientific notation.
   b. Convert scientific notation to decimal numbers.
   c. Add, subtract, multiply, and divide numbers expressed in scientific notation (Section 2.5).
6. You should be able to express any number as a logarithm or convert a logarithm to a natural number (Section 2.6).

7. You should be able to plot and interpret line graphs of two variables that are functions of each other (Section 2.7).
8. You should know the common Metric and SI prefixes for decimal multiples and fractions (Section 2.8).
9. You should know the common units of measurement for length, volume, and mass in the International and Metric Systems (Section 2.8).
10. You should know approximate comparisons between the Metric or SI units of mass, length, and volume to similar units of the U.S. Customary System (Section 2.10).
11. You should be able to set up solutions for simple word problems by means of dimensional-unit analysis (Section 2.9).
12. You should be able to use conversion factors and dimensional-unit analysis to convert a given measurement from the U.S. Customary System to the Metric System or vice versa (Section 2.10).
13. You should know the relationship between common U.S. Customary units and metric units.
14. You should be familiar with the common tools of measurement of mass and volume (Section 2.11).
15. When you record a measurement, you should know how many significant figures to use (Section 2.11).
16. You should be able to determine the allowable number of significant figures for an answer in arithmetical operations using significant figures (Section 2.11).
17. You should be able to convert temperature readings between the Fahrenheit, Celsius, and Kelvin scales (Section 2.12).
18. You should know the meaning of density (density is the ratio of the mass of a body to its volume) and be able to express the definition as an algebraic equation. (density $= \frac{mass}{volume}$) (Section 2.13).
19. Given any two of the measurements, density, mass, or volume, you should be able to determine the third. (Remember to use dimensional-unit analysis to set up or check your solution.) (Section 2.13)

IMPORTANT TERMS AND CONCEPTS

SECTION 2.4

Exponents   the number placed as a superscript following a base number which indicates the number of times the base number is to be multiplied by itself or in the case of a fractional exponent the root the base number is to be reduced to.

Section 2.5

Scientific notation   the convention of expressing a number as a decimal number between 1 and 10 multiplied by 10 raised to an exponential power to give the appropriate magnitude.

Section 2.6

Logarithms   the exponential power to which a base number (commonly 10) must be raised to be equal to a given number.

Section 2.10

Conversion factor   the ratio of two equivalent quantities of measure having different units that is equal to unity (1).

Section 2.12

Temperature   the condition of a body which determines the transfer of heat to or from another body.

Absolute zero   the theoretical temperature at which all molecular motion ceases. It is assigned the value of 0 K or -273.15°C.

Section 2.13

Density   the ratio of the mass of a body of matter to the volume of the body.

Specific gravity   the ratio of the mass of a body of matter to the mass of an equal volume of a standard substance, usually water for liquids and solids.

QUESTIONS AND PROBLEMS

Math Review

1. Perform the following indicated addition or subtraction of fractions:

a. $\dfrac{5}{9} + \dfrac{2}{3} =$ b. $\dfrac{2}{3} + \dfrac{2}{7} =$ c. $\dfrac{5}{2} + \dfrac{3}{4} + \dfrac{2}{3} =$

d. $\dfrac{1}{2} + \dfrac{5}{6} + \dfrac{3}{7} + \dfrac{2}{3} =$ e. $\dfrac{a}{b} + \dfrac{c}{d} =$ f. $\dfrac{a}{b} + \dfrac{b}{a} =$

g. $\dfrac{2}{3} - \dfrac{1}{2} =$ h. $\dfrac{4}{5} - \dfrac{2}{3} =$ i. $\dfrac{4x}{y} - \dfrac{2}{z} =$

j. $\dfrac{a}{c} - \dfrac{a}{b} =$

Answers:

1. To add or subtract fractions, they are converted to fractions with a common denominator and only the numerators are added or subtracted.

   a. The common denominator of 9 and 3 is 9

   $$\frac{5}{9} + \frac{2}{3} = \frac{5}{9} + \frac{6}{9} = \frac{11}{9}$$

   b. $\dfrac{20}{21}$

   c. The common denominator of 2, 4, and 3 is 12

   $$\frac{5}{2} + \frac{3}{4} + \frac{2}{3} = \frac{30}{12} + \frac{9}{12} + \frac{8}{12} = \frac{47}{12}$$

   d. $\dfrac{102}{42}$

   e. $\dfrac{a}{b} + \dfrac{c}{d} = \dfrac{ad}{bd} + \dfrac{cb}{bd} = \dfrac{ad + bc}{bd}$

   f. $\dfrac{a^2 + b^2}{ab}$

   g. $\dfrac{2}{3} - \dfrac{1}{2} = \dfrac{4}{6} - \dfrac{3}{6} = \dfrac{1}{6}$ h. $\dfrac{2}{15}$

   i. $\dfrac{4x}{y} - \dfrac{2}{z} = \dfrac{4xz}{yz} - \dfrac{2y}{yz} = \dfrac{4xz - 2y}{yz}$ j. $\dfrac{ab - ac}{cb}$

2. Multiply the following fractions as indicated:

   a. $\dfrac{1}{3} \times \dfrac{3}{4} =$ b. $\dfrac{5}{7} \times \dfrac{3}{8} =$ d. $\dfrac{a}{b} \times \dfrac{b}{c} =$

   e. $\dfrac{a}{d} \times \dfrac{2d}{b} \times \dfrac{b}{3c} =$

Answers:

2. a. $\dfrac{1}{\cancel{3}} \times \dfrac{\cancel{3}}{4} = \dfrac{1}{4}$ b. $\dfrac{5}{7} \times \dfrac{3}{8} = \dfrac{15}{56}$ d. $\dfrac{a}{\cancel{b}} \times \dfrac{\cancel{b}}{c} = \dfrac{a}{c}$

e. $\dfrac{a}{\cancel{d}} \times \dfrac{2\cancel{d}}{\cancel{b}} \times \dfrac{\cancel{b}}{3c} = \dfrac{2a}{3c}$

(Note: When there are identical terms in the numerator and the denominator of a series of multiplications, they cancel. This is an important procedure in using dimensional-unit analysis or the factor-label method of solving problems.)

3. Perform the indicated division of fractions:

a. $\dfrac{3}{4} \div \dfrac{2}{3} =$ b. $\dfrac{7}{12} \div \dfrac{7}{9} =$

c. $\dfrac{4x}{5y} \div \dfrac{x}{5} =$ d. $a \div \dfrac{4a}{c} =$

Answers:

3. a. $\dfrac{3}{4} \div \dfrac{2}{3} = \dfrac{3}{4} \times \dfrac{3}{2} = \dfrac{9}{8}$ b. $\dfrac{7}{12} \div \dfrac{7}{9} = \dfrac{7}{12} \times \dfrac{9}{7} = \dfrac{9}{12} = \dfrac{3}{4}$

c. $\dfrac{4x}{5y} \div \dfrac{x}{5} = \dfrac{4\cancel{x}}{\cancel{5}y} \times \dfrac{\cancel{5}}{\cancel{x}} = \dfrac{4}{y}$ d. $a \div \dfrac{4a}{c} = \cancel{a} \times \dfrac{c}{4\cancel{a}} = \dfrac{c}{4}$

SECTION 2.1

Solve the following problems using a scientific pocket calculator.

4. a. 2175 b.   375 c. 2656
       3962   4695   8912
       8431   1237   3726
       5257   3841   6996
       ————   ————   ————

d. 1479 - 765 + 5927 - 3747 =

e. 8492 - 6557 + 7483 - 4176 =

f. 5280 ÷ 941 x 375 =

g. 84 479 x 245 ÷ 362 + 419 =

Answers:

a. 19 825   b. 10 148   c. 22 290   d. 2894
e. 5242   f. 2140   g. 136.4

SECTION 2.2   Negative Numbers

5. Perform the following additions:

  a. $5 + (-2) + (+6) + (-10) =$

  b. $-273 + (-40) + (-60) =$

  c. $4x + 4 + (2x - 2) =$

  d. $5 + (-2) + x + (-6x) =$

6. Perform the following subtractions:

  a. $-40 - (-273) =$ b. $-10 - (-32) =$

  c. $-7x - (-3x) =$ d. $(6x + 10) - (4x - 16) =$

7. Perform the indicated multiplications and divisions:

  a. $(-2)(+6)(-4) =$ b. $(-x)(-4)(-5) =$

  c. $(4x)(-y)(x - y) =$ d. $144 \div (-6) =$

Answers:

5. In the addition of negative numbers: (i) if all of the numbers are negative the sum is negative; (ii) if the signs of the two numbers are different, the difference is taken and the sign of the larger number is retained.

  a. -1   b. -373   c. $6x + 2$ d. $3 - 5x$

6. In the subtraction of negative numbers, the sign of the number is changed and the two numbers are added algebraically.

  a. $-40 - (-273) = -40 + 273 = 233$

  b. $-10 - (-32) = -10 + 32 = 22$ c. $-4x$

  d. $(6x + 10) - (4x - 16) = 6x + 10 - 4x + 16 = 2x + 26$

7. In multiplication and division, if an even number of negative numbers are multiplied or divided the product or the quotient is positive; if an odd number of negative numbers are multiplied or divided, the product or quotient is negative.

   a. +48   b. -20x c. $-4x^2y + 4xy^2$ d. -24

SECTION 2.3 Solving Equations

8. Solve: $xy + a = 7$ for x

9. Solve: $d = \dfrac{m}{v}$ for v

Answers:

8. Isolate the term containing x on the left of the equation by subtracting a from each side:

$$
\begin{array}{rcl}
xy + a &=& 7 \\
\underline{\quad - a} & & \underline{\quad - a} \\
xy &=& 7 - a
\end{array}
$$

Divide both sides of the equation by y

$$\frac{x\cancel{y}}{\cancel{y}} = \frac{7 - a}{y}$$

$$x = \frac{7 - a}{y}$$

9. $d = \dfrac{m}{v}$

Multiply both sides by v

$$d \times v = \frac{m}{\cancel{v}} \times \cancel{v}$$

$$d \times v = m$$

Divide both sides by d

$$\frac{\cancel{d} \times v}{\cancel{d}} = \frac{m}{d}$$

$$v = \frac{m}{d}$$

Section 2.4 Exponents

10. Write the following in exponential form:

a. $(2)(2)(2)(2)(2) =$ b. $\frac{1}{7} \times \frac{1}{7} \times \frac{1}{7} \times \frac{1}{7} =$

c. $(y)(y)(y)(y)(y)(y)(y) =$

11. Solve the following:

a. 6^5 b. 2^{-6} c. $81^{1/4}$

d. $36^{3/2}$ e. 10^{-3} f. 8^0

12. Solve the following multiplications and divisions:

a. $x^3 \cdot x^2 =$ b. $(y^4)(y^{-6}) =$

c. $(y^2)(y^2)(y^3) =$ d. $x^3 \div x^2 =$

e. $y^{-4} \div y^3 =$ f. $m^{-5} \div m^{-8} =$

13. Solve the following:

a. $(y^3)^2 =$ b. $(x^8)^{1/4} =$ c. $(y^3)^{4/3} =$ d. $(y)^0$

Answers:

10. a. 2^5, a number times itself a series of times may be written as the number followed by a superscript showing the number of times it is to be multiplied by itself; the superscript is called an exponent.

b. 7^{-4}, a fraction written in an exponential form may be written as the reciprocal of the fraction to the negative power.

c. y^7

11. a. $6 \times 6 \times 6 \times 6 \times 6 = 7776$

b. $2^{-6} = \frac{1}{2} \times \frac{1}{2} \times \frac{1}{2} \times \frac{1}{2} \times \frac{1}{2} \times \frac{1}{2} = \frac{1}{64}$

c. $81^{1/4} = \sqrt[4]{81} = 3$, a number to a fractional power is the number to the root of the demoninator.

d. $36^{3/2} = (\sqrt{36})^3 = (6)^3 = 216$

e. $10^{-3} = \dfrac{1}{1000} = 0.001$

f. $8^0 = 1$, any number to the zero power is equal to 1.

12. a. x^5 , when exponential numbers are multiplied, the exponents are added.

b. y^{-2} c. y^7

d. x , when exponential numbers are divided, the exponent of the divisor is subtracted from the exponent of the dividend.

e. y^{-7} f. m^3

13. a. y^6 , when an exponential number is raised to a power, the exponent of the number is multiplied by the power.

b. x^2 c. y^4

d. $y^0 = 1$, any number to the 0 power is equal to 1.

SECTION 2.5   Scientific Notation

14. Express the following numbers in scientific notation:

a. 1 576 000   b. 3 724.6

c. $825 000 000   d. 2 470 000 000

e. 5 280   f. 224 000

g. 0.000 293   h. 0.000 000 006 75

i. 0.017   j. 0.000 049 6

k. 0.046 75   l. 0.197

15. Convert the following scientific notations to decimal numbers:

   a. 5.89×10^4 b. 3.950×10^{-3}

   c. 2.067×10^7 d. 8.14×10^{-5}

   e. 7.756×10^{-2} f. 4.29×10^{16}

16. Multiply or divide the following numbers as indicated:

   a. $(6.02 \times 10^{23})(2.15 \times 10^{-5})$

   b. $(5.42 \times 10^6) \div (6.02 \times 10^{23})$

   c. $(2.01 \times 10^{-23})(1.20 \times 10^{24})$

   d. $(7.68 \times 10^{-10}) \div (3.45 \times 10^{-19})$

Answers:

14. a. To have only one digit to the left of the decimal, the decimal must be moved 6 places to the left. $1\,576\,000 = 1.576 \times 10^6$

   b. Move the decimal 3 places to the left: $3\,724.6 = 3.724\,6 \times 10^3$

   c. Move the decimal 8 places to the left: $\$825\,000\,000 = \$8.25 \times 10^8$

   d. 2.47×10^9 e. 5.28×10^3

   f. 2.24×10^5

   g. The decimal must be moved 4 places to the right. The exponent of 10 will be -4.
        $0.000\,293 \times 10^{-4}$

   h. The decimal must be moved 9 places to the right.
        $0.000\,000\,006\,75 = 6.75 \times 10^{-9}$

i. The decimal must be moved 2 places to the right.
   $0.017 = 1.7 \times 10^{-2}$

j. 4.96×10^{-5} k. 4.675×10^{-2}

l. 1.97×10^{-1}

15. a. $5.89 \times 10^4 = 5.89 \times 10\ 000 = 58\ 900$

    b. $3.950 \times 10^{-3} = 3.950 \times 0.001 = 0.003\ 950$

    c. $2.067 \times 10^7 = 2.067 \times 10\ 000\ 000 = 20\ 670\ 000$

    d. $0.000\ 0814$ e. 0.07756

    f. $42\ 900\ 000\ 000\ 000\ 000$

16. Multiply or divide the decimal number and the expo-
    nential powers of 10 separately. Then convert the
    result to correct scientific notation if necessary.

    a. $(6.02 \times 10^{23})(2.15 \times 10^{-5}) = 12.9 \times 10^{18} = 1.29 \times 10^{19}$

    b. 9.00×10^{-18} c. 2.41×10

    d. 2.23×10^9

SECTION 2.6 Logarithms

17. Using a scientific pocket calculator, convert the
    following numbers to logarithms:

    a. $1\ 760$ b. $0.021\ 48$

    c. 6.72×10^{-11} d. 6.02×10^{23}

    e. 2.41×10^{-5} f. 1.76×10^{-4}

18. Convert the following logs to scientific notation:

    a. 8.5939 b. -5.48 c. 3.475

Answers:

17. a. 3.246 b. -1.668 c. -10.173

d. 23.780   e. -4.618   f. -3.754

18. a. 3.9255×10^{8} b. 3.31×10^{-6}

c. 2.985×10^{3}

SECTION 2.7   Graphs

19. Using the pressure vs altitude graph Figure 2.2 of the textbook, determine the following:

    a. the pressure at an altitude of 2.5 km.
    b. the altitude at which the pressure is 278 torr.
    c. the pressure at an altitude of 11.0 km.

Answers:

19. a. Find the altitude of 2.5 km along the horizontal axis. Draw a vertical line from this point to intersect the pressure vs altitude plot. Draw a horizontal line from the point of intersection to the pressure axis. Estimate the value of the pressure. The pressure is 559 torr.

    b. Find the value of 278 torr on the pressure axis. Draw a horizontal line to intersect with the pressure vs altitude plot. Draw a vertical line from this point of intersection to the altitude axis. The altitude is 7.2 km.

    c. Use the same procedure as a. The pressure is 170 torr.

SECTION 2.8   The International System of Measurements

20. Give the International System of Measurements standard unit for each of the following measurements:

    a. length   b. mass   c. volume

21. Give the Metric measurement standard unit for each of the following:

    a. length   b. mass   c. volume

22. Give the multiple for each of the following SI or metric prefixes:

    a. mega- (M)
    b. kilo- (k)
    c. deci- (d)
    d. centi- (c)
    e. milli- (m)
    f. micro- (μ)

23. Express 5 g in units of μg, mg, and kg.

Answers:

20. a. length = meter   b. mass = kilogram, kg
    c. volume = cubic meter, m^3

21. a. length = meter   b. mass = kilogram, kg
    c. volume = liter

22. a. mega- (M) = 1 000 000 or 10^6
    b. kilo- (k) = 1 000 or 10^3
    c. deci- (d) = 0.1 or 10^{-1}
    d. centi- (c) = 0.01 or 10^{-2}
    e. milli- (m) = 0.001 or 10^{-3}
    f. micro- (μ) = 0.000 001 or 10^{-6}

23. 5 g = 5 000 000 g = 5 000 mg = 0.005 kg

SECTION 2.9 Solving Problems with Dimensional Analysis

24. Water running from a pipe fills a 10 gallon can in 250 seconds. What is the rate of flow of water in the pipe in gal/sec?

25. If the rate of flow of water in a river is 300 ft^3/min, how long will it take for 285 000 ft^3 to flow past a point?

Answers:

24. Given: quantity = 10 gal; time = 250 sec
    Find: rate = gal/sec
    From the units given for rate, we can see that we must divide the quantity, gal, by the time, sec, to get gal/sec

$$\text{rate} = \frac{10 \text{ gal}}{250 \text{ sec}} = 0.04 \text{ gal/sec}$$

25. <u>Given</u>: volume = 285 000 ft^3; rate = 300 ft^3/min
    <u>Find</u>: time = min

    Since the answer must have min in the numerator, we must divide by the rate.

    $$\frac{\text{ft}^3}{\text{ft}^3/\text{min}} = \cancel{\text{ft}^3} \times \frac{\text{min}}{\cancel{\text{ft}^3}} = \text{min}$$

    $$\frac{285\,000 \text{ ft}^3}{300 \text{ ft}^3/\text{min}} = 285\,000 \cancel{\text{ft}^3} \times \frac{1 \text{ min}}{300 \cancel{\text{ft}^3}} = 950 \text{ min}$$

SECTION 2.10   Conversion of Units of Measurement

26. What is a conversion factor?

27. From Table 2.6 in the textbook, write the conversion factors required for each of the following conversions:

    a. meters to yards   b. quarts to liters
    c. kilometers to miles   d. pounds to kilograms

28. How many liters are there in 27 ft^3?
    (1 m^3 = 35.3 ft^3)

29. If gasoline costs $1.57/gal, what is the cost per liter?   (1 m^3 = 264.3 gal)

30. If a pitcher throws a fast ball 90 miles per hour, what is its speed in meters per second?

<u>Answers</u>:

26. A conversion factor is the ratio of two equivalent measurements that have different units that is equal to unity or 1.

27. Note: the conversion factor should always have the units that you are converting from in the denominator.

a. $\dfrac{1.094 \text{ yd}}{1 \text{ m}}$ b. $\dfrac{1 \text{ liter}}{1.057 \text{ qt}}$

c. $\dfrac{0.621 \text{ mi}}{1 \text{ km}}$ d. $\dfrac{1 \text{ kg}}{2.205 \text{ lb}}$

28. **Given:** 27 ft^3
    $\quad$ 1 m^3 = 35.3 ft^3
    $\quad$ 1 m^3 = 1 000 liter
    Find: liters

$$27 \text{ ft}^3 \times \frac{1 \text{ m}^3}{35.3 \text{ ft}^3} \times \frac{1 \text{ 000 liter}}{1 \text{ m}^3} = 765 \text{ liters}$$

29. **Given:** $1.57/gal
    $\quad$ 1 m^3 = 264.3 gal
    $\quad$ 1 m^3 = 1 000 liter
    Find: $/liter

$$\frac{\$1.57}{\text{gal}} \times \frac{264.3 \text{ gal}}{1 \text{ m}^3} \times \frac{1 \text{ m}^3}{1 \text{ 000 liter}} = \frac{\$0.415}{\text{liter}}$$

30. **Given:** 90 mi/hr
    $\quad$ 0.62 mi = 1 km
    $\quad$ 1 km = 1 000 m
    $\quad$ 60 sec = 1 min
    $\quad$ 60 min = 1 hr
    Find: m/sec

$$90 \frac{\text{mi}}{\text{hr}} \times \frac{1 \text{ km}}{0.62 \text{ mi}} \times \frac{1 \text{ 000 m}}{1 \text{ km}} \times \frac{1 \text{ hr}}{60 \text{ min}} \times \frac{1 \text{ min}}{60 \text{ sec}} = 40 \frac{\text{m}}{\text{sec}}$$

SECTION 2.11   Uncertainty of Measurements and
$\qquad\qquad$ Significant Figures

31. How many significant figures are there in each of
    the following numbers?

    a. 1.450   b. 2 360   c. 3 062
    d. 20 010   e. 0.002   f. 0.005 010

Answers:

31. a. 4 , the zero is significant since it is not
    otherwise needed.

b. 3 , the zero is not significant since its purpose is to locate the decimal.

c. 4 , any zero between two other numbers is significant.

d. 4 , the two zeroes between the 2 and the 1 are
significant; the zero on the right of the 1 is
merely a place holder.

e. 1 , none of the zeroes are significant since
they are merely locating the decimal point.

f. 4 , the zeroes to the left of the 5 are not significant but the one to the right of the 1 is.

SECTION 2.12   Temperature Scale

32. On a summer day when the temperature is $96^{\circ}F$, what
is the reading on the Celsius scale?

33. What is absolute zero on the Fahrenheit scale?

34. How do you convert from the Kelvin scale to the
Celsius scale?

Answers:

32. Given: $f^{\circ}F = 96^{\circ}F$
Find: $c^{\circ}C$

$$1.8c + 32 = f$$
$$1.8c + 32 = 96$$
$$1.8c = 64$$
$$c = 36$$
$$c^{\circ}C = 36^{\circ}C$$

33. Given: Absolute zero $= -273.15^{\circ}C$
Find: Absolute zero $= f^{\circ}F$

$$1.8c + 32 = f$$
$$1.8(-273.15) + 32 = f$$
$$-426.87 + 32 = f$$

$$-394.87 = f$$

Absolute zero $= -394.87^\circ F$

34. Subtract 273 from the Kelvin reading.

SECTION 2.13 Density and Specific Gravity

35. Find the density of a substance if 12.40 cm^3 of the substance has a mass of 84.82 g.

36. If a liquid has a density of 1.424 g/cm^3, what volume of liquid must be measured to have 109.0 g?

37. What is the mass of 75.00 cm^3 of a liquid that has a density of 1.347 g/cm^3?

Answers:

35. Given: mass = 84.82 g; volume = 12.40 cm^3
    Find: density = g/cm^3

    Since the density has the units g/cm^3, we simply need to divide the mass by the volume.

    $$density = \frac{84.82\,g}{12.40\ cm^3} = 6.840\ \frac{g}{cm^3}$$

36. Given: mass = 109.0 g; density = 1.424 $\frac{g}{cm^3}$
    Find: volume = cm^3

    The dimensional units show us that we must divide the mass by the density.

    $$volume = \frac{109.0\ g}{1.424\ \frac{g}{cm^3}} = \frac{109.0\ g}{1.424} \times \frac{cm^3}{g} = 76.54\ cm^3$$

37. Given: volume = 75.00 cm^3; density = 1.347 $\frac{g}{cm^3}$
    Find: mass = g

    Dimensional-unit analysis shows us that

    $$\frac{g}{cm^3} \times cm^3 = g$$

    $$mass = 1.347\ \frac{g}{cm^3} \times 75.00\ cm^3 = 101.0\ g$$

SELF-TEST   1. Solve the equation: $PV = nRT$ for n.

2. Express the following in scientific notation:
   a. 4 460 000   b. 0.000 035 16

3. Convert the following scientific notations to decimal numbers:
   a. 1.819×10^{-5} b. 1.76×10^{11}

4. Solve the following:
   a. $9.61 \times 10^5 + 1.57 \times 10^4 + 7.60 \times 10^3 =$

   b. $(6.49 \times 10^5)(3.85 \times 10^6)(5.42 \times 10^{-7}) =$

   c. $(1.75 \times 10^{-4})(3.96 \times 10^{-2}) \div (6.84 \times 10^{-8}) =$

5. a. $\log 7.71 \times 10^{-10} =$
   b. $\log 6.02 \times 10^{23} =$

6. Find the number whose logarithm is:
   a. -7.477   b. 3.691

7. Convert the following masses to grams: (Use only 1.000 kg = 2.204 lb, and 1.00 lb = 16.00 oz. Do not refer to tables to convert from one metric unit to another.)
   a. 2.342 kg   b. 7 160 mg
   c. 120 lb   d. 4.50 oz

8. Convert the following volumes to liters: (Use only 1 m^3 = 35.3 ft^3, 1 liter = 1.057 qt, 1 gal = 4 qt; do not refer to tables for the metric prefixes.)
   a. 2500 ml   b. 5.00 ft^3 c. 10 gal

9. Determine the number of significant digits in each of the following numbers:
   a. 42 309   b. 0.002 070
   c. 635 000   d. 10.03

10. Make the following conversions:
    a. -40°C to °F   b. 392°F to °C   c. 5.0 K to °C

11. If 10.00 cm^3 of a liquid weighs 12.79 g what is its density?

12. What is the mass of a cubic block of metal measuring 2.50 cm on each side if its density is 10.5 g/cm^3?

13. What volume of a liquid with a density of 1.017 g/cm^3 is required in order to have 213 g of substance?

Answers:

1. $n = \dfrac{PV}{RT}$

2. a. 4.46×10^6 b. 3.516×10^{-5}

3. a. 0.000 018 19   b. 176 000 000 000

4. a. 9.84×10^5 b. 1.35×10^6
   c. 1.01×10^2

5. a. -9.113   b. 23.780

6. a. 3.33×10^{-8} b. 4.91×10^3

7. a. 2 342 g   b. 7.160 g
   c. 54 450 g   d. 128 g

8. a. 2.5 liters   b. 142 liters
   c. 38 liters

9. a. 5   b. 4   c. 3   d. 4

10. a. -40°F   b. 200°C
    c. -268.2°C

11. 1.279 $\dfrac{g}{cm^3}$

12. 164 g

13. 209 cm^3

chapter

3 Matter & Energy

OBJECTIVES
1. After completing your study of this chapter, you should be able to give precise definitions for the terms associated with matter and the classification of matter (Sections 3.1 and 3.2).
2. You should be able to distinguish between the classes of matter and the properties of the various forms of matter (Section 3.2).
3. You should be familiar with the representation of elements by the use of symbols (Section 3.3).
4. You should understand the Law of Definite Composition and the Law of Multiple Proportion and why compounds always have fixed compositions (Section 3.5).
5. You should be able to give precise definitions for the terms: solid, liquid, gas, melting point, freezing point, and boiling point (Sections 3.6 and 3.7).
6. You should know the meaning of the terms chemical change and physical change (Sections 3.4 and 3.6).
7. You should know the Law of Conservation of Mass and Energy (Section 3.10).

IMPORTANT TERMS AND CONCEPTS

SECTION 3.1

Matter   anything that has mass and occupies space.

Mass   the manifestation of the property of inertia.

Inertia   the tendency of a body to remain at rest if at rest or to resist changes in motion until it is acted on by an outside force.

Property   any distinguishing quality, characteristic
    attribute, or typical mode of action which belongs
    or pertains to a form of matter.

Extrinsic properties   properties common to all forms
    of matter

Intrinsic properties   properties which are unique to
    one type of matter distinguishing it from other
    types or forms of matter.

SECTION 3.2

Homogeneous   uniform throughout in composition and in-
    trinsic properties.

Phase   a single body of homogeneous matter.

Substance   a homogeneous form of matter with a fixed
    composition.

SECTION 3.3

Element   a substance that cannot be broken down into
    similar substances by ordinary chemical means. Of
    the 106 currently known elements, 91 are naturally
    occurring. At least 24 elements are contained in
    those substances that are essential to life.

Atom   the smallest particle of an element.

SECTION 3.4

Compound   a substance composed of two or more elements
    which are chemically combined.

Molecule   the smallest particle of a compound. All
    molecules of the same compound have identical atomic
    compositions; that is, they contain exactly the same
    kind and the same number of atoms.

Chemical change   the transformation of one or more
    substances into one or more different substances.
    In a chemical change the composition of substances
    always changes.

SECTION 3.5

The Law of Definite Composition   the law states that
    when two or more elements combine to form a given

compound, they always combine in a fixed ratio by weight.

The Law of Multiple Proportion   the law states that when the same elements combine to form more than one compound, the different weights of one of the elements that combine with a fixed weight of another are in the ratio of small whole numbers. (What this means is that if X and Y combine to form two compounds and a fixed weight of Y is used in each case, then the ratio of the weight of X in compound A to the weight of X in compound B will be 1:2, 1:3, 2:3, or some similar ratio in small numbers.)

Section 3.6

Physical states   the mode of existence determined by intrinsic properties and external condition (solid, liquid, and gas).

Solid   the physical state of matter in which the particles of a substance (atoms or molecules) are held in fixed positions relative to one another by strong cohesive forces between particles so that a definite shape and volume are maintained.

Liquid   the physical state of matter in which the particles are relatively free to change position with one another but are held together by cohesive forces strong enough to maintain a fixed volume.

Gas   the physical state of matter in which particles are practically unrestricted by cohesive forces so that neither a fixed volume nor a fixed shape are retained.

SUMMARY OF THE PHYSICAL STATES OF MATTER

state	volume	shape
solid	fixed	fixed
liquid	fixed	unrestricted
gas	unrestricted	unrestricted

Section 3.7

Physical change   a change in the state of a substance but not in the composition of the substance.

Melting point or freezing point   the temperature at which a substance undergoes a transition from the solid state to the liquid state or the liquid state to the solid state.

Boiling point   the temperature at which a liquid undergoes a transition to a gas at atmospheric pressure with turbulent bubbling. (You will learn a more rigorous definition of boiling point in Chapter 9.)

SECTION 3.8

Crude mixture   a combination of two or more discontinuous phases.

Solution   a homogeneous form of matter that has a variable composition.

SECTION 3.9

Energy   the capability of doing work.

Potential energy   energy a body of matter possesses because of its composition, state, or position.

Kinetic energy   energy a body of matter possesses becauses of its motion.

SECTION 3.10

Conservation of matter   in ordinary chemical reactions matter can be transformed but it cannot be created or destroyed.

Conservation of energy   energy may be converted from one form to another but it cannot be created or destroyed.

Conservation of mass and energy   the total quantity of mass and energy of the universe is fixed. The law for the interconversion of mass and energy is

$$E = mc^2$$

SECTION 3.11

System   that part of the universe that is under study.

Surroundings   that part of the universe that is exclusive of the system.

QUESTIONS
AND
PROBLEMS

SECTION 3.3   Elements

1. Given the atomic number and the name of the following elements what is the symbol for each element?

Atomic Number	Name	Symbol		Atomic Number	Name	Symbol
1	hydrogen	____		11	sodium	____
2	helium	____		12	magnesium	____
3	lithium	____		13	aluminum	____
4	beryllium	____		14	silicon	____
5	boron	____		15	phosphorus	____
6	carbon	____		16	sulfur	____
7	nitrogen	____		17	chlorine	____
8	oxygen	____		18	argon	____
9	fluorine	____		19	potassium	____
10	neon	____		20	calcium	____

2. Give the name for each of the following elements.

Symbol	Name		Symbol	Name
Na	_____		O	_____
Be	_____		B	_____
F	_____		Cl	_____
S	_____		I	_____
N	_____		Ar	_____
Li	_____		C	_____
H	_____		Br	_____

Symbol	Name	Symbol	Name
Ne	_____	He	_____
Si	_____	P	_____
Mg	_____	K	_____
Al	_____	Ca	_____

Check your answers for Questions 1 and 2 against a list of the elements. The first 20 elements are so important that it may be wise to memorize the list in Question 1.

SECTION 3.5 Laws of Composition

3. Sulfur dioxide is composed of 50% by weight sulfur and 50% by weight oxygen. What weight of oxygen is required to react with 16 g of sulfur?

4. Exactly 2.70 g of aluminum combine with 2.40 g of oxygen to form aluminum oxide. What weight of oxygen is required to react with 4.05 g of aluminum?

5. A sample of an oxide of phosphorus was found to contain 3.10 g of phosphorus and 2.40 g of oxygen. Another sample of an oxide of phosphorus was found to contain 1.55 g of phosphorus and 2.00 g of oxygen.
   a. Were the two samples samples of the same compound?
   b. What law of composition is illustrated by the two samples?

Answers:

3. The Law of Definite Composition states that elements form compounds in a fixed ratio by weights. A 2 g sample of sulfur dioxide would contain 1 g of sulfur (50% by weight) and 1 g oxygen.

$$\frac{x \text{ g oxygen}}{16 \text{ g S}} = \frac{1 \text{ g oxygen}}{1 \text{ g S}}$$

$$x \text{ g oxygen} = \frac{1 \text{ g oxygen}}{1 \text{ g S}} \times 16 \text{ g S}$$

$$x \text{ g oxygen} = 16 \text{ g oxygen}$$

4.
$$\frac{x \text{ g oxygen}}{4.05 \text{ g Al}} = \frac{2.40 \text{ g oxygen}}{2.70 \text{ g Al}}$$

$$x \text{ g oxygen} = \frac{2.40 \text{ g oxygen}}{2.70 \text{ g Al}} \times 4.05 \text{ g Al}$$

$$x \text{ g oxygen} = 3.60 \text{ g oxygen}$$

5. a. 1st sample: $\dfrac{3.10 \text{ g P}}{2.40 \text{ g O}} = \dfrac{1.29 \text{ g P}}{1 \text{ g O}}$

   2nd sample: $\dfrac{1.55 \text{ g P}}{2.00 \text{ g O}} = \dfrac{0.775 \text{ g P}}{1 \text{ g O}}$

   They are not the same compound since the ratio of the weights of the elements are not the same.

   b. The two compounds illustrate the Law of Multiple Proportion since the ratio of the weight of P that reacts with 1 g of oxygen in the first compound to the weight of P that reacts with 1 g of oxygen in the second compound is 5:3.

$$\frac{1.29}{0.775} = \frac{1.66}{1} = \frac{5}{3}$$

SECTION 3.10   Conservation of Matter and Energy

6. When butane reacts with oxygen, carbon dioxide and water are produced. If 4.40 g of butane react with 16.00 g of oxygen and the weight of the carbon dioxide produced is 13.20 g, what is the weight of water that was produced?

Answers:

6. Conservation of matter requires that the mass of the reactants always equals the mass of the products. Therefore, if the total mass of the reactants is 4.40 g + 16.00 g = 20.40 g

The mass of the products must be 20.40 g.
The weight of water produced is 7.20 g.

SELF-TEST
1. The extensive property of matter that accounts for mass is _____.
2. Density, color, physical state at a given temperature, and the boiling point are all _____ of a substance.
3. In order to be classified as a substance, matter must be _____ and have _____.
4. The two classes of substances are _____ and _____.
5. Homogeneous matter that has a variable composition is classified as _____.
6. The three states of matter are _____, _____, and _____.
7. A body of matter that consists of only one phase is said to be _____.
8. A change in a substance that does not involve any change in composition is called a _____ change.
9. The capacity to do work is called _____.
10. A heterogeneous or crude mixture must consist of two or more _____.
11. Any time a substance has a change in composition a _____ change must have occurred.
12. The smallest particle of an element is _____.
13. The temperature at which a substance changes from the solid state to the liquid state is called _____.

Answers:

1. inertia
2. intrinsic properties
3. must be *homogeneous* and have *a constant composition*.
4. elements and compounds
5. solutions
6. solid, liquid, and gas
7. homogeneous
8. physical change

9. energy
10. phases
11. chemical change
12. atom
13. melting point

chapter

4 Atomic Structure

OBJECTIVES
1. After completing this chapter, you should be able to explain the periodic behavior of the elements (Section 4.1).
2. You should be able to list the three major subatomic particles and the relative size and electrical charge of each (Section 4.2).
3. You should be able to state Dalton's atomic theory in your own words and be able to tell how it differed from the modern concept of the atom (Section 4.3).
4. You should be able to explain the modern concept of the atom and give evidence that supports it (Section 4.3).
5. You should be able to explain the meaning of atomic numbers (Section 4.4).
6. You should be able to explain how the masses of the various atoms of elements are assigned (Section 4.5).
7. You should be able to explain the meaning of isotopes (Section 4.6).
8. You should be able to describe the general characteristics of an atomic nucleus (Section 4.7).
9. You should be able to recognize the symbol for isotopes of the various elements (Section 4.6).
10. You should be able to determine the number of electrons, protons, and neutrons in the atoms of various isotopes from their atomic number and mass number (Section 4.6).

31

11. You should be able to explain the meaning of natural radioactivity, unstable nuclei, and the zone of stability (Section 4.8, optional).

12. You should be able to list the four types of nuclear decay and to write equations for each kind of decay (Section 4.8, optional).

13. You should know the distinction between atomic mass and atomic weight (Section 4.9).

14. You should be able to calculate the atomic weight of an element from the natural abundance and the atomic mass of its various isotopes (Section 4.9).

15. You should be able to list the four parameters that describe the behavior of the electrons in an atom and the significance of each of these parameters (Sections 4.11 - 4.14).

16. You should know the number of subshells in each main shell, the number of orbitals in each subshell, and the number of electrons in each orbital, subshell, and shell (Section 4.11 - 4.14).

17. You should be able to write the spectroscopic notation for any atom from the atomic number of the element (Section 4.15).

18. You should be able to draw electron configuration and energy level diagrams for the atoms of the elements from their atomic numbers (Section 4.16).

IMPORTANT TERMS AND CONCEPTS

SECTION 4.1

<u>Periodic classification of the elements</u>   a chart of all the elements in which they are arranged in groups according to an ordered repetition of their physical and chemical properties.

<u>Cathode ray tube</u>   the name given a gas discharge tube that shows rays emanating at the cathode and directed toward the anode. Cathode rays are streams of electrons.

<u>Electron</u>   a subatomic particle that has a mass of 9.109×10^{-28} g (0.00055 amu or 1/1850 the mass of a hydrogen atom) and a negative electrical charge of 4.8×10^{-10} electrostatic units. (The relative electrical charge of an electron is -1.)

<u>Proton</u>   a subatomic particle with a mass of 1.0073 amu or 1.6726×10^{-24} g and a charge equal in magnitude but opposite in sign to an electron. (Protons are

usually assigned a charge of +1.)

Neutron   a subatomic particle with the approximate
    mass of a proton but without an electrical charge.

SUMMARY OF SUBATOMIC PARTICLES

particle	symbol	mass (amu)	relative charge	location in the atom
electron	e^-	0.00055	-1	electron cloud
proton	H^+ or p^+	1.0073	+1	nucleus
neutron	n	1.0087	0	nucleus

SECTION 4.3

Dalton's Atomic Theory
  a. Matter is composed of distinct miniscule parti-
     cles (atoms) that can neither be subdivided nor
     broken down into simpler substances.
  b. Atoms are indestructible.
  c. All atoms of the same element are identical.
  d. Atoms of different elements have different
     weights.
  e. When elements combine to form compounds, the
     atoms of two or more elements react with one an-
     other in ratios of small whole numbers, such as
     1:1, 1:2, 1:3, 2:3, 2:5, etc.
  f. The relative weights of elements that combine
     with each other, according to the Law of Definite
     Composition, are proportional to the relative
     weights of the atoms of the elements.

Rutherford's Atom   the modern concept of the atom
    which regards it as consisting of an extremely small
    nucleus that comprises almost all of the mass and a
    diffuse electron cloud that accounts for its volume.

SECTION 4.4

Atomic number   the number assigned to an atom which
    designates the number of protons in the nucleus and
    the number of electrons in the electron cloud. The
    atomic number is written as a subscript preceding

the symbol for the element: $_1H$, $_2He$, $_3Li$, $_4Be$, etc.

Section 4.5

Atomic mass   the mass of an atom based on the mass of
    the most common isotope of carbon which is assigned
    a mass of exactly 12. The exact masses of the atoms
    of the various isotopes of each element are deter-
    mined using a mass spectrometer (Figure 4.9 in the
    text). The exact masses of the various isotopes of
    the elements are very close to whole numbers; the
    whole number is called the mass number. For exam-
    ple: magnesium consists of three isotopes, mass
    23.98504, mass 24.98584, and mass 25.98259. These
    isotopes are assigned mass numbers of 24, 25, and
    26 respectively.

Section 4.6

Isotopes   atoms of the same element that have differ-
    ent masses. Atoms of isotopes have the same numbers
    of protons and electrons but different numbers of
    neutrons. The atoms of isotopes are identified by
    using the notation:

$$_Z^A E$$

where E is the symbol of the element, z is the
    atomic number of the element (z = number of protons
    = number of electrons), and A is the mass number of
    the isotope (A = number of protons + number of neu-
    trons; $A - z$ = number of neutrons).

Section 4.7

Nucleus   the part of the atom that is made up of pro-
    tons and neutrons. The nucleus is extremely small
    (radius ~ 10^{-13} to 10^{-12} cm).

Section 4.8 (optional)

Radioactivity   the process of decay of unstable nuclei
    accompanied by the emission of small particles from
    the nucleus and high-energy electromagnetic radia-
    tion (gamma rays).

Beta decay   nuclear decay in which a beta particle is
    emitted from the nucleus.

Beta <u>particle</u> (β^-)   a high-energy electron emitted
   from the nucleus.
<u>Positron</u> <u>decay</u>   nuclear decay in which a positron is
   emitted from the nucleus.
<u>Positron</u> (β^+)   a high-energy subatomic particle equal
   in mass but opposite in charge to an electron.
<u>Electron</u> <u>capture</u>   radioactive decay in which the nu-
   cleus captures an electron from the first energy
   level of the electron cloud.
<u>Alpha</u> <u>particle</u> <u>emission</u>   the most common radioactive
   decay for elements with high atomic numbers in which
   the neutron to proton ratio is too high involving
   the emission of a high-energy alpha particle from
   the nucleus.
<u>Alpha</u> <u>particle</u> ($_2^4He^{2+}$)   a high energy helium nucleus.
<u>Gamma</u> <u>rays</u>   extremely high-energy electromagnetic
   radiation that is emitted during nuclear decay.

Section 4.9

<u>Atomic</u> <u>weight</u>   the weighted average of the natural
   abundance of isotopes of an element times the atomic
   mass of each isotope.

Section 4.10

Practically the entire volume of the atom is made up of
   the electron cloud. The velocity of an electron in
   the electron cloud is approximately 0.1 times the
   speed of light.

Section 4.11

<u>Principal</u> <u>quantum</u> <u>number</u> (n)   a constant derived from
   quantum mechanics which denotes the main energy
   level (shell) of the electron in an atom and has the
   value of any integer 1, 2, 3, etc.
<u>Azimuthal</u> <u>quantum</u> <u>number</u> (l)   a constant derived from
   quantum mechanics which denotes the energy sublevel
   (subshell) of the electron in the atom and has the
   values of 0, 1, 2, etc. to $n - 1$.
<u>Magnetic</u> <u>quantum</u> <u>number</u> (m)   a constant derived from
   quantum mechanics which denotes the orbital of the
   energy sublevel (subshell) of the electron in an
   atom and has the values of $-l$ through 0 to $+l$.
<u>Spin</u> <u>quantum</u> <u>number</u> (s)   the number from quantum

mechanics that denotes the direction of apparent
spin of the electron in the atom. The values of the
spin quantum number are either +1/2 or -1/2.

SECTION 4.12

The main energy level is frequently called the shell
and can hold up to a maximum of $2n^2$ electrons (n is
the value of the principal quantum number). The
main energy levels in an atom are designated as 1,
2, 3, etc.

SECTION 4.13

The number of subshells in each shell is the same as
the numerical value of the principal quantum number,
i.e.:

$n = 1$	1st shell	1 subshell
$n = 2$	2nd shell	2 subshells
$n = 3$	3rd shell	3 subshells

The subshells are designated s, p, d, f, g, h, etc.

SECTION 4.14

Orbital   a region in space around an atomic nucleus in
which there is a high probability of electron density. An orbital can contain no more than two electrons and then only if they have opposing spins.

a. All s subshells have one orbital.
b. All p subshells have three orbitals.
c. All d subshells have five orbitals.
d. All f subshells have seven orbitals.

Pauli's exclusion principle   no two electrons in the
same atom can have a set of identical quantum numbers.

SECTION 4.15

Aufbau theory   Aufbau simply means buildup and it is
the building up of electronic structure of the atoms
based on the quantum-mechanical solution of simple
atoms. In the spectroscopic notation used to show
electronic structure:

a. The main shell is indicated by the appropriate Arabic numeral.

b. The subshells are designated by s, p, d, f.

c. The number of electrons in each subshell is shown by an Arabic numeral written as a superscript following the notation for the subshell. $3p^5$ indicates 5 electrons in the p subshell of the third main shell.

QUESTIONS AND PROBLEMS

SECTION 4.2 Subatomic Particles

1. List the three most important subatomic particles, give their relative charge, approximate mass, and location in the atom.

SECTIONS 4.4, 4.5, AND 4.6 Atomic Number, Atomic Mass, and Isotopes

2. Indicate the number of electrons, protons, and neutrons in each of the following atoms:

a. $_2^3$He

b. $_{27}^{59}$Co

c. $_3^6$Li

d. $_{31}^{69}$Ga

e. $_{34}^{80}$Se

f. $_8^{18}$O

3. Indicate similarities and differences in the following atoms:

a. $_8^{16}$O and $_8^{18}$O

b. $_{16}^{36}$S and $_{18}^{36}$Ar

c. $_{22}^{50}$Ti and $_{24}^{52}$Cr

d. $_{28}^{58}$Ni and $_{28}^{64}$Ni

e. $_{31}^{69}$Ga and $_{32}^{70}$Ge

f. $_{54}^{134}$Xe and $_{56}^{134}$Ba

Answers:

1.

particle	charge	approximate mass	location in the atom
electron	-1	1/1850 amu	electron cloud
proton	+1	1 amu	nucleus
neutron	0	1 amu	nucleus

2. a. $_2^3$He, the atomic number 2 indicates the number of electrons and protons, the mass number is the sum of protons and neutrons. The difference in the mass number and the atomic number gives the number of neutrons: 3 - 2 = 1 neutron.

   2 electrons, 2 protons, and 1 neutron

   b. $_{27}^{59}$Co   atomic number 27, mass number 59
   27 electrons, 27 protons, and 32 neutrons

   c. $_3^6$Li   atomic number 3, mass number 6
   3 electrons, 3 protons, and 3 neutrons

   d. $_{31}^{69}$Ga   atomic number 31, mass number 69
   31 electrons, 31 protons, and 38 neutrons

   e. $_{34}^{80}$Se   atomic number 34, mass number 80
   34 electrons, 34 protons, and 46 neutrons

   f. $_8^{18}$O   atomic number 8, mass number 18
   8 electrons, 8 protons, and 10 neutrons

3. a. Both atoms have the same atomic number, therefore, they are isotopes of the same element. They have the same number of electrons and protons but different numbers of neutrons.
   b. Both atoms have the same mass numbers, but different atomic numbers. They have approximately the same mass but different numbers of protons, electrons, and neutrons.
   c. Both atoms have different atomic numbers and different mass numbers, but the difference between the respective mass numbers and atomic numbers is 28 in each case. Therefore, the two atoms have the same number of neutrons in their respective nuclei.
   d. The atomic number is the same for both of these atoms; therefore, they are isotopes of the same element with the same number of protons and electrons but different numbers of neutrons.
   e. Both atoms have different mass numbers and different atomic numbers; however, the difference in mass number and atomic number for each is 38. Therefore, the only similarity is that each of

them has 38 neutrons.

f. Both of these atoms have the same mass number, therefore, their atomic masses are similar. However, since the atomic numbers are different, they have different numbers of protons, electrons, and neutrons.

SECTION 4.8 (OPTIONAL) Natural Radioactivity

4. Fill in the blanks for the following reactions. Indicate which type of radioactive decay is occurring.

a. $^{31}_{14}Si \rightarrow$ _____ $+ \, ^{0}_{-1}e$

b. $^{14}_{6}C \rightarrow \, ^{14}_{7}N +$ _____

c. _____ $\rightarrow \, ^{16}_{8}O + \, ^{0}_{-1}e$

d. $^{14}_{8}O \rightarrow$ _____ $+ \, ^{0}_{+1}e$

e. $^{36}_{17}Cl + \, ^{0}_{-1}e \rightarrow$ _____

f. $^{226}_{88}Ra \rightarrow \, ^{222}_{86}Rn +$ _____

g. $^{211}_{83}Bi \rightarrow$ _____ $+ \, ^{4}_{2}He$

Answers:

4. a. $^{31}_{15}P$, beta decay

b. $^{0}_{-1}e$, beta decay

c. $^{16}_{7}N$, beta decay

d. $^{14}_{7}N$, positron decay

e. $^{36}_{16}S$, electron capture

f. $^{4}_{2}He$, alpha particle emission

g. $^{207}_{81}Tl$, alpha particle emission

SECTION 4.9 Atomic Weights

5. Calculate the atomic weights of the following

elements from the given natural abundance and atomic
mass of their isotopes:

a. Sb	isotope	natural abundance	mass
	$^{121}_{51}Sb$	57.25%	120.9
	$^{123}_{51}Sb$	42.75%	122.9

b. Ga	isotope	natural abundance	mass
	$^{69}_{31}Ga$	60.40%	68.93
	$^{71}_{31}Ga$	39.60%	70.92

c. Si	isotope	natural abundance	mass
	$^{28}_{14}Si$	92.21%	27.98
	$^{29}_{14}Si$	4.700%	28.98
	$^{30}_{14}Si$	3.090%	29.97

Answers:

5. a. 0.5725 x 120.9 = 69.22
      0.4275 x 122.9 = 52.53
       atomic weight = 121.75

   b. 0.6040 x 68.93 = 41.63
      0.3960 x 70.92 = 28.08
       atomic weight = 69.71

   c. 0.9221 x 27.98 = 25.80
      0.0470 x 28.98 =   1.36
      0.0309 x 29.97 =   0.93
       atomic weight = 28.09

SECTIONS 4.10 - 4.16 Electronic Configuration of Atoms

6. Give the maximum number of electrons in each of the
   following shells:
   a. 1   b. 3   c. 5

7. Give the maximum number of electrons in each of the

subshells:
a. *s*   b. *p*   c. *d*   d. *f*

8. If a *g* subshell were filled with electrons, how many electrons would it hold?

9. Give the number of electrons in each main shell of the following atoms:
   a. $_8O$ b. $_{53}I$ c. $_{33}As$ d. $_{20}Ca$

10. Write the spectroscopic notation for the electronic structure of each of the following atoms:
   a. $_2He$ b. $_{10}Ne$ c. $_{15}P$ d. $_{28}Ni$
   e. $_{48}Cd$ f. $_{82}Pb$

Answers:

6. The maximum number of electrons that can populate a given shell is $2n^2$.
   a. 1st shell $2n^2 = 2 \times 1^2 = 2$
   b. 3rd shell $2n^2 = 2 \times 3^2 = 18$
   c. 5th shell $2n^2 = 2 \times 5^2 = 50$ (Note: The 5th shell could have a *g* subshell if there were enough electrons in any atom to fill it.)

7. a. The *s* subshell has one orbital and can contain 2 electrons.
   b. The *p* subshell has three orbitals and can contain 6 electrons.
   c. The *d* subshell has five orbitals and can contain 10 electrons.
   d. The *f* subshell has seven orbitals and can contain 14 electrons.

8. A *g* subshell would have nine orbitals and could contain 18 electrons.

9. a. Oxygen: 1st shell 2
              2nd shell 4

   b. Iodine: 1st shell 2
              2nd shell 8
              3rd shell 18
              4th shell 18
              5th shell 7

  c. Arsenic: 1st shell 2
              2nd shell 8
              3rd shell 18
              4th shell 5

  d. Calcium: 1st shell 2
              2nd shell 8
              3rd shell 8
              4th shell 2

10. a. $_2$He , $1s^2$

   b. $_{10}$Ne , $1s^2\, 2s^2\, 2p^6$

   c. $_{15}$P , $1s^2\, 2s^2\, 2p^6\, 3s^2\, 3p^3$

   d. $_{28}$Ni , $1s^2\, 2s^2\, 2p^6\, 3s^2\, 3p^6\, 4s^2\, 3d^8$

   e. $_{48}$Cd , $1s^2\, 2s^2\, 2p^6\, 3s^2\, 3p^6\, 4s^2\, 3d^{10}\, 4p^6\, 5s^2\, 4d^{10}$

   f. $_{82}$Pb , $1s^2\, 2s^2\, 2p^6\, 3s^2\, 3p^6\, 4s^2\, 3d^{10}\, 4p^6\, 5s^2\, 4d^{10}\, 5p^6$
       $6s^2\, 4f^{14}\, 5d^{10}\, 6p^2$

SELF-TEST
1. The _____ accounts for practically all of the volume of the atom.
2. Practically all of the mass of the atom is contained in the _____.
3. The subatomic particle with a mass of about 0.0005 amu and a charge of -1 is called _____.
4. One of the major subatomic particles that has no electrical charge is the _____.
5. The atomic number indicates the number of _____ and _____ in an atom.
6. A region in space around an atomic nucleus in which there is a high probability of electron density is called _____.
7. The atoms $^{16}_{8}$O and $^{18}_{8}$O are called _____.
8. The major difference between $^{16}_{8}$O and $^{18}_{8}$O is the number of _____ in each.
9. An atom represented as $^{37}_{17}$Cl has _____ electrons, _____ protons, and _____ neutrons.
10. The transformation of $^{42}_{18}$Ar to $^{42}_{19}$K is an example of

_____ decay.

11. The transformation of $^{190}_{78}$Pt to $^{186}_{76}$Os is an example of _____ decay.

12. Positron emission by $^{50}_{25}$Mn will result in a nuclide with an atomic number of _____ and a mass number of _____.

13. A type of radioactivity that only removes excess energy from the nucleus is called _____.

14. An element is composed of two isotopes; one has a natural abundance of 57.25% and a mass of 120.9 amu, the other has a natural abundance of 42.75% and a mass of 122.9 amu. What is the atomic weight of the element? _____

15. Atoms of the element $_{17}$Cl have electrons in how many main energy levels or shells? _____

16. A p subshell can contain a maximum of _____ electrons.

17. There are _____ orbitals in a d subshell.

18. The four parameters or conditions that describe an electron in an atom are: _____, _____, _____, and _____.

19. The spectroscopic notation for the electronic structure of $_{33}$As is _____.

20. The atomic number of the element whose atoms have an electronic structure of $1s^2\,2s^2\,2p^6\,3s^2\,3p^6\,4s^2\,3d^5$ is _____.

Answers:

1. electron cloud
2. nucleus
3. an electron
4. neutron
5. electrons and protons
6. an orbital
7. isotopes
8. neutrons
9. 17 electrons, 17 protons, and 20 neutrons
10. beta
11. alpha particle
12. 24 and 50
13. gamma radiation
14. 121.75
15. 3

16. 6
17. 5
18. the main energy level or shell, the energy sublevel or subshell, the orbital, and the spin
19. $1s^2\, 2s^2\, 2p^6\, 3s^2\, 3p^6\, 4s^2\, 3d^{10}\, 4p^3$
20. 25

chapter

5 The Periodic Table

OBJECTIVES
1. You should be able to identify the position of an element in the Periodic Table from either its symbol or its atomic number (Section 5.1).
2. You should be able to recognize the value of the atomic weight of an element as it is usually shown in a Periodic Table (Section 5.1).
3. You should be able to recognize the sections of the Periodic Table that relate to the various subshells of the electronic structure of the atoms (Section 5.1).
4. You should be able to relate the position of the individual elements in the Periodic Table to the electronic structure of their atoms (Section 5.2).
5. You should be able to identify the subshell that the last electron to be added to an atom of a particular element occupies (Section 5.2).
6. You should be able to determine the number of valence electrons that atoms of each element have (Section 5.3).
7. You should be able to identify those elements that belong to each particular group or family and understand why they belong to that group or family of elements (Section 5.4).
8. You should understand the meaning and significance of a noble-gas configuration (Section 5.9).
9. You should know the meaning of the terms ionization potential, electron affinity, electronegativity, and atomic size and be able to explain the significance

45

of each of these terms (Sections 5.7 - 5.10).

10. From the position of an element in the Periodic Table, you should be able to give a comparison of the magnitude of the size, ionization potential, electron affinity, and electronegativity of its atoms relative to those of the atoms of other elements (Sections 5.7 - 5.10).

11. You should be able to draw the Lewis structure of the atoms of the various elements (Section 5.6).

12. From its position in the Periodic Table, you should be able to identify an element as a metal or a non-metal (Section 5.11).

IMPORTANT TERMS AND CONCEPTS

SECTION 5.1

Periodic Table can be considered an inverted energy-level diagram of the electronic structure of the atoms of the elements. The Periodic Table is divided into four major sections representing the s, the p, the d, or the f subshells. (See Figure 5.3 of the text.)

SECTION 5.2

Since the Periodic Table can be regarded as a modified energy-level diagram for the electron configuration of the atoms of the elements, the electron configuration of the atoms of each of the elements may be determined from the position of the element in the Periodic Table. (See the problems for this section on the following pages.)

SECTION 5.3

Valence electrons the electrons that occupy the outermost s and p orbitals. Valence electrons are responsible for the chemical properties of each element.

SECTION 5.4

Elements whose atoms have similar electronic configurations for their valence electrons have similar chemical properties.

Group or <u>family</u> <u>of</u> <u>elements</u>   those elements that are
in the vertical columns in the Periodic Table. Ele-
ments in the same group or family have similar electronic configurations for their valence electrons
and therefore have similar chemical properties.

<u>Alkali</u> <u>metal</u> <u>family</u> (group IA)   consists of $_3$Li, $_{11}$Na,
$_{19}$K, $_{37}$Rb, and $_{87}$Fr, the atoms of which each have 1
electron in their outermost shell which is in an s
subshell.

<u>Alkaline</u> <u>earth</u> <u>family</u> (group IIA)   consists of $_4$Be,
$_{12}$Mg, $_{20}$Ca, $_{38}$Sr, $_{56}$Ba, and $_{88}$Ra which have two
electrons in the outermost s subshell as their val-
ence electrons.

<u>Halogen</u> <u>family</u> (group VIIA)   consists of $_9$F, $_{17}$Cl,
$_{35}$Br, $_{53}$I, and $_{85}$At which have outer shells with the
configuration of ns^2np^5 for a total of seven valence
electrons.

Group <u>VIA</u>   family of elements consists of $_8$O, $_{16}$S,
$_{34}$Se, $_{52}$Te, and $_{84}$Po in which the atoms' outermost
s and p subshells contain 2 and 4 electrons, respectively, for a total of six valence electrons.

Group <u>VA</u>   family of elements is composed of $_7$N, $_{15}$P,
$_{33}$As, $_{51}$Sb, and $_{83}$Bi in which the atoms' outermost
s and p subshells contain 2 and 3 electrons, respectively, for a total of five valence electrons.

Group <u>IVA</u>   family of elements is composed of $_6$C, $_{14}$Si,
$_{32}$Ge, $_{50}$Sn, and $_{82}$Pb in which the atoms' outermost
s and p subshells contain 2 and 2 electrons, respectively, for a total of four valence electrons.

Group <u>IIIA</u>   family of elements consists of $_5$B, $_{13}$Al,
$_{31}$Ga, $_{49}$In, and $_{81}$Tl which each have atoms with an
outer shell electron configuration of ns^2np^1 for a
total of three valence electrons.

SECTION 5.5

<u>Transition</u> <u>metal</u> <u>elements</u>   atomic number 21-30, 39-
48, and 71-80, are those elements in which the var-
ious d subshells of the atoms are filling. In each
case the respective d subshell is interior to the
next higher s subshell which is already populated
with electrons.

<u>Inner</u> <u>transition</u> <u>metal</u> <u>elements</u>   consist of those ele-
ments in which the $4f$ subshell is filling (the

lanthanide series, atomic numbers 58 - 70) or the 5*f* subshell is filling (the actinide series, atomic numbers 89 - 102).

SECTION 5.6

<u>Lewis</u> <u>structure</u>   of an atom consists of the symbol for the element with dots representing the valence electrons of the atom.

SECTION 5.7

<u>Atomic</u> <u>size</u> or <u>atomic</u> <u>radius</u>   the size or radius of the atoms of the elements. Atomic size decreases for the elements as the position moves from left to right across a given period of the Periodic Table and increases as the position of the element moves from top to bottom of a group of elements in the Periodic Table.

SECTION 5.8

<u>Ionization</u> <u>potential</u>   the energy required to remove an electron from an isolated atom of an element. Ionization potentials generally increase from left to right across a given period and decrease from top to bottom in a given group of elements in the Periodic Table.

SECTION 5.9

<u>Electron</u> <u>affinity</u>   the energy change that accompanies the addition of an electron to an isolated atom of an element to form a negative ion. Electron affinities are generally the highest for the halogens and decrease from right to left moving away from the halogens; they also generally decrease moving from the top of a group to the bottom of a group in the Periodic Table. The noble gases He, Ne, Ar, Kr, and Rn have extremely low electron affinities which indicates the stability of a filled outer *s* and *p* subshell.

SECTION 5.10

<u>Electronegativity</u>   the attraction an atom has for a

pair of electrons shared with another atom. Electronegativity is highest for fluorine and decreases
with the increase in distance that an element is
from fluorine in the Periodic Table.

SECTION 5.11

Metals   those elements that are to the left of B, Si,
  As, Te, and Po in the Periodic Table.
  Properties of metals are:
  a. low ionization potentials,
  b. low electron affinities,
  c. low electronegativities
  d. good to high conductivity of both heat and elec-
     tricity,
  e. malleability, the property that lets them be ham-
     mered and rolled into shapes without breaking.

Nonmetals   those elements to the right of B, Si, As,
  Te, and Po in the Periodic Table.
  Properties of nonmetals are:
  a. high ionization potentials,
  b. high electron affinities
  c. medium to high electronegativities,
  d. generally poor conductivity of both heat and
     electricity,
  e. nonmetal solids are usually brittle and easily
     pulverized.

QUESTIONS
AND
PROBLEMS

SECTION 5.1 Sections of the Periodic Table

1. From the Periodic Table list the elements in which
   the $2p$ subshell is being filled.
2. From the Periodic Table list the elements in which
   the $5s$ subshell of the atoms is being filled.
3. From the Periodic Table list the elements in which
   the $4d$ subshell of the atoms is being filled.

Answers:

1. Atomic numbers 5 - 10. B, N, C, O, F, and Ne.
2. 37Rb and 38Sr.
3. Atomic numbers 39 - 48: Y, Zr, Nb, Mo, Tc, Ru, Rh,
   Pd, Ag, and Cd.

SECTION 5.2   Periodic Classification and Electronic
              Structure

4. Give the atomic number and symbol for the element
   whose atoms have the following electronic configuration:
   a. $1s^2$
   b. $1s^2\,2s^2$
   c. $1s^2\,2s^2\,2p^3$
   d. $1s^2\,2s^2\,2p^6\,3s^2$
   e. $1s^2\,2s^2\,2p^6\,3s^2\,3p^3$
   f. $1s^2\,2s^2\,2p^6\,3s^2\,3p^6\,4s^2$
   g. $1s^2\,2s^2\,2p^6\,3s^2\,3p^6\,4s^2\,3d^5$
   h. $1s^2\,2s^2\,2p^6\,3s^2\,3p^6\,4s^2\,3d^{10}\,4p^3$

5. Using the Periodic Table give the number of electrons in each of the following subshell of the atoms
   of the designated element.
   a. The $2s$ subshell of $_3$Li.
   b. The $2p$ subshell of $_6$C.
   c. The $2p$ subshell of $_{11}$Na.
   d. The $5p$ subshell of $_{54}$Xe.
   e. The $4f$ subshell of $_{80}$Hg.
   f. The $5p$ subshell of $_{52}$Te.
   g. The $4d$ subshell of $_{39}$Y.
   h. The $5d$ subshell of $_{80}$Hg.

6. Give the number of electrons in the designated shell
   of the atoms of the following elements.
   a. The 2nd shell of $_8$O.
   b. The 3rd shell of $_{32}$Ge.
   c. The 5th shell of $_{53}$I.
   d. The 4th shell of $_{82}$Pb.
   e. The 5th shell of $_{54}$Xe.

Answers:

4. We can determine the atomic number of the element by
   adding the superscripts that stand for the number of
   electrons in each shell. We can find the element
   either from its atomic number or from its electronic
   configuration in the outer shell.
   a. From the Periodic Table we can see that $_2$He is
      the element with a filled 1st shell.
   b. The element that has atoms with 2 electrons in

the $2s$ subshell is $_4$Be.
  c. The element that has atoms with 3 electrons in
      the $3p$ subshell is $_7$N.
  d. The element that has atoms with 2 electrons in
      the $3s$ subshell is $_{12}$Mg.
  e. $_{15}$P
  f. $_{20}$Ca
  g. $_{25}$Mn
  h. $_{33}$As

5. a. In the Periodic Table, the $2s$ subshell is repre-
       sented by the first two squares in the 2nd period
       or horizontal row. Li is the first element in
       the row so it has 1 electron in the $2s$ subshell.
   b. The last 6 squares in the 2nd period represent
       the $2p$ subshell; C is the 3rd square so it has 3
       electrons in the $2p$ subshell.
   c. Na is in the 3rd period in which the $3s$ subshell
       of the atoms are filling; therefore the $2p$ sub-
       shell is already filled with 6 electrons.
   d. The $5p$ subshell is represented by the last 6
       squares of the 5th period; therefore Xe has 6
       electrons in the $5p$ subshell.
   e. The $4f$ subshell is filling with elements atomic
       numbers 58 - 71; Hg is element 80; therefore the
       $4f$ subshell is filled with 14 electrons.
   f. 1 electron
   g. 1 electron
   h. 10 electrons

6. a. Oxygen is the 6th element in the 2nd period;
       therefore it has a configuration of $2s^2 2p^4$ for a
       total of 6 electrons in the 2nd shell.
   b. Ge is in the 4th period; the $3s$ and $3p$ subshells
       are filled after the 3rd period and the $3d$ sub-
       shell is filled atomic numbers 21 - 30; therefore
       Ge has $3s^2 3p^6 3d^{10}$ or 18 electrons in the 3rd
       shell.
   c. I is in the 5th period; atomic numbers 37 - 38 re-
       present the two $5s$ electrons, atomic numbers
       39 - 48 the ten $4d$ electrons, and atomic numbers
       49 - 53 five of the $5p$; therefore I has 7 elec-
       trons in the 5th shell.
   d. Atomic numbers 19 - 20 represent $4s^2$, 31 - 36 re-
       present $4p^6$, atomic numbers 39 - 40 the $4d^{10}$ elec-
       trons, and atomic numbers 58 - 71 the $4f^{14}$;

tnerefore $_{82}$Pb has a completely filled 4th shell
with 32 electrons.
e. 8 electrons

SECTIONS 5.3 AND 5.4   Valence Shell Electrons and
Groups or Families of Elements

7. Explain what is meant by a group or family of elements.

8. List two elements that have the same number of valence electrons as $_{14}$Si.

9. List two other elements that are in the same family of elements as $_{35}$Br.

10. Using the spectroscopic notation, list the valence shell electrons for the group IIIA elements.

11. List those elements that have a valence shell with electrons with the configuration of $ns^2 np^6$.

Answers:

7. A group or family of elements consists of those elements that have similar electronic configurations and the same number of valence electrons. These elements are always in the vertical columns of the Periodic Table.

8. The elements that have the same number of valence electrons are in the vertical columns of the Periodic Table. For Si these would be C, Ge, Sn, or Pb.

9. Bromine is a member of the halogen family with F, Cl, I, and At.

10. The group IIIA elements have the valence shell configuration of $ns^2 np^1$ and include B, $2s^2 2p^1$; Al, $3s^2 3p^1$; Ga, $4s^2 4p^1$; In, $5s^2 5p^1$; and Tl, $6s^2 6p^1$.

11. The valence shell configuration of $ns^2 np^6$ denotes a filled valence shell or the noble gas family: Ne, Ar, Kr, Xe, and Rn.

SECTION 5.6   Drawing Lewis Structures of the Atoms

12. Draw the Lewis representation for an atom of each
of the following elements:
a. Cs   d. As   g. Xe
b. Si   e. Ba   h. Se
c. I   f. In

Answers:

12. a. Cs has 1 valence electron; its Lewis structure
    is: Cs·
    b. Si has 4 valence electrons; its Lewis structure
    is: S·̈·
    c. I has 7 valence electrons: :Ï:
    d. As has 5 valence electrons: ·Äs·
    e. Ba has 2 valence electrons: Ba:
    f. In has 3 valence electrons: Ïn·
    g. Xe has 8 valence electrons: :Xë:
    h. Se has 6 valence electrons: ·Së·

SECTIONS 5.7 - 5.10   Atomic Size, Ionization Potentials,
                       Electron Affinities, and Electro-
                       negativity

13. Account for the following facts:
    a. Nitrogen atoms are smaller than boron atoms.
    b. Germanium, Ge, atoms are larger than silicon,
       Si, atoms.
    c. The noble gas elements have high 1st ionization
       potentials.
    d. The difference between the 1st and 2nd ioniza-
       tion potentials for group IA elements is much
       greater than the difference between the 1st and
       2nd ionization potentials of the group IIA ele-
       ments.
    e. The 1st ionization potential for potassium, K,
       is smaller than the 1st ionization potential for
       sodium, Na.

14. Using the Periodic Table, make the following

comparisons:
a. The electronegativity of oxygen atoms to fluorine atoms.
b. The size of sulfur atoms to chlorine atoms.
c. The 1st ionization potential of Br compared to Kr.
d. The difference between the 1st and 2nd ionization potentials of Ca and the difference between the 2nd and 3rd ionization potentials of Ca.
e. The electron affinity of Cl compared to the electron affinity of S.
f. The electron affinity of Ar compared to the electron affinity of Cl.

15. Classify the following elements as metals or non-metals:

a. B	g. As	m. Bi
b. Be	h. Ne	n. I
c. Na	i. H	o. Zn
d. Al	j. N	p. Hg
e. S	k. Sb	
f. Sn	l. P	

Answers:

13. a. Nitrogen atoms are smaller than boron atoms because an increase in the nuclear charge without an expansion of the number of shells results in a stronger attraction for the electrons in each shell causing a contraction in size.
b. Germanium atoms are larger than silicon atoms because the screening effect of the inner shells of electrons results in less attraction for electrons in shells more distant from the nucleus.
c. The noble gas elements have high 1st ionization potentials because the outer configuration of $ns^2 np^6$ or a filled valence shell is extremely stable.
d. The 2nd ionization of the group IA elements requires that an electron be removed from an ion with a noble gas configuration while the 2nd ionization of the group IIA elements simply removes the 2nd electron from the valence shell which results in a noble gas configuration.

e. the 1st ionization potential for K is less than that for Na because each of the valence electrons of the respective atoms is attracted by a net +1 kernel charge but the electron in the K **atoms** is further from the nucleus and, therefore, the attraction is less.

14. a. Fluorine is the most electronegative of all atoms.
   b. Atomic size decreases from left to right across a period; therfore, S atoms are larger than Cl atoms.
   c. Kr, a noble gas, has a filled valence shell and therefore has the highest ionization potential of any of the elements in the same period.
   d. The removal of the 3rd electron from a Ca atom requires intrusion into the filled $3s^2 3p^6$ shell; therefore the difference between the 1st and 2nd ionization potentials is less than the difference between the 2nd and 3rd ionization potentials.
   e. Cl has the greater electron affinity.
   f. Ar is a noble gas with a stable configuration and has a minimal electron affinity; Cl has a high electron affinity.

15. Metals: b. Be   c. Na   d. Al   f. Sn   k. Sb
    m. Bi   o. Zn   and p. Hg
    Nonmetals: a. B   e. S   g. As   h. Ne   i. H
    j. N   l. P   and n. I

SELF-TEST   1. From its position in the Periodic Table, one would predict that the last electron to pass to:
   a. an iron, Fe, atom is in the _____ subshell.
   b. a bromine, Br, atom is in the _____ subshell.
   c. a calcium, Ca, atom is in the _____ subshell.
   d. a europium, Eu, atom is in the _____ subshell.
   e. a gallium, Ga, atom is in the _____ subshell.

2. Using a Periodic Table, determine the elements whose atoms have electronic configurations ending as shown:
   Example: $...4s^2 3d^{10} 4p^6$ Answer: Kr
   a. $...2s^2 2p^3$
   b. $...3s^2 3p^5$

c. $\ldots 4s^2\, 3d^5$
d. $\ldots 6s^2\, 5d^5$
e. $\ldots 6p^3$
f. $\ldots 7s^2$
g. $\ldots 4d^{10}$
h. $\ldots 5p^6$
i. $\ldots 5d^6$
j. $\ldots 6p^4$

3. Fill in the blanks.
   a. There are _____ electrons in the 3d subshell of
      $_{32}$Ge.
   b. $_{15}$P atoms have _____ valence electrons.
   c. There are _____ electrons in the 4p subshell of
      $_{34}$Se.
   d. The element with the smallest atoms that belongs
      to the same family as Br is _____.
   e. The element that has the smallest atoms that is
      in the same family as potassium, K, is _____.
   f. The element that has the largest atoms of any of
      the elements in the 4th period is _____.
   g. Oxygen, fluorine, sulfur, and chlorine are clas-
      sified as _____, while sodium, magnesium,
      potassium, calcium, and zinc are classified as
      _____.
   h. Of the 3rd period elements, _____ has the highest
      ionization potential.
   i. _____ atoms are the most electronegative of any
      of the elements.
   j. Of all of the 3rd period elements, _____ has the
      highest electron affinity.

4. Draw the Lewis structures for atoms of the following
   elements:
   a. Na   c. Br   e. Ne
   b. Al   d. S

Answers:

1. a. 3d c. 4s e. 4p
   b. 4p d. 4f

2. a. $_7$N   e. $_{83}$Bi   i. $_{76}$Os
   b. $_{17}$Cl   f. $_{88}$Ra   j. $_{84}$Po
   c. $_{25}$Mn   g. $_{48}$Cd
   d. $_{75}$Re   h. $_{54}$Xe

3. a. 10
   b. 5
   c. 4
   d. F, fluorine
   e. Li, lithium

   f. K, potassium
   g. nonmetals,...metals
   h. Ar, argon
   i. F, fluorine
   j. Cl, chlorine

4. a. Na
   b. Al
   c. Br

   d. S
   e. Ne

chapter

6 Chemical Bonds Between Atoms

OBJECTIVES
1. You should be able to explain what is meant by an ionic or electrovalent bond (Section 6.1).
2. You should be able to explain what is meant by a covalent bond (Section 6.1).
3. You should be able to explain what is meant by the "octet rule" (Section 6.2).
4. You should be able to explain how metal cations are formed from metal atoms and how nonmetal anions are formed from nonmetal atoms (Section 6.3).
5. You should be able to explain the variations in size of ions from the size of their parent atoms (Section 6.3).
6. You should be able to list some of the properties that characterize a compound as an ionic substance (Section 6.4).
7. You should be able to draw Lewis structures for both positive and negative ions (Section 6.5).
8. You should be able to predict and write the formula of any ionic binary compound (Section 6.6).
9. You should understand the importance of covalent bonding (Section 6.7).
10. You should be able to explain what molecules are and how they are formed (Section 6.8).
11. You should be able to list some of the properties of covalent compounds (Section 6.9).
12. You should be able to draw Lewis structures for simple covalent molecules (Section 6.10).
13. You should be able to compare and contrast the

characteristics of ionic compounds and covalent compounds (Sections 6.4 and 6.10, Table 6.21).

14. You should be able to predict and to write the formulas of simple covalent compounds using the Periodic Table (Section 6.11).

15. You should be able to predict the geometry of covalent molecules using the electron repulsion theory (Section 6.12).

16. You should be able to predict the polarity of covalent bonds based on your knowledge of the variation of the electronegativity of atoms with the position of the elements in the Periodic Table (Section 6.13).

17. Using your knowledge of molecular geometry and polarity of covalent bonds, you should be able to predict the polarity of molecules (Section 6.14).

18. You should be able to explain what is meant by a coordinate covalent bond (Section 6.15).

19. You should be able to recognize the common polyatomic ions from their formulas and be able to give their names and ionic charge and to write the formulas of the common polyatomic ions from their names (Section 6.16).

20. You should be able to determine the formal charge on an ion from the Lewis structure of the polyatomic ion (Section 6.17).

21. Using the charge of metal ions predicted from the Periodic Table and the charge of polyatomic ions, you should be able to write the formulas of ionic compounds containing the common polyatomic ions (Section 6.18).

22. You should be able to predict the charge of the less common polyatomic ions from the formulas of their compounds with monatomic ions whose charge you can predict from the Periodic Table (Section 6.19).

23. You should be able to describe what is meant by a metallic bond (Section 6.20).

24. Referring to the rules of nomenclature in Section 6.21, you should be able to name most simple inorganic compounds (Section 6.21).

IMPORTANT TERMS AND CONCEPTS

SECTION 6.1

Ionic bond   the strong electrostatic force between positive and negative ions in a compound.

Covalent compound   compounds in which the atoms are bonded by a sharing of pairs of electrons.

SECTION 6.2

Octet rule   the rule of bonding that states that atoms gain stability if they can acquire an octet or eight electrons in their valence shell and therefore have a noble gas configuration.

Isoelectronic structures   atomic particles with the same electronic configurations.

SECTION 6.3

Cations   atoms or aggregates of atoms (molecules) that have fewer electrons than the neutral atoms or molecules and consequently have a positive charge.

Anions   atoms or aggregates of atoms (molecules) that have more electrons than the neutral atoms or molecules and consequently have a negative charge.

Size of monatomic positive ions   it is always smaller than the size of the neutral atom. The size of monatomic positive ions decreases with an increase in positive charge.

Size of monatomic negative ions   it is always greater than the size of the neutral atom. Size of monatomic negative ions increases with an increase in ionic charge.

SECTION 6.4

Properties of ionic compounds include:
    a. High melting points (usually $300^{\circ}C$).
    b. High electrical conductivity in the molten state.
    c. Good electrical conductivity when dissolved in water.

SECTION 6.5

The Lewis structures of monatomic positive ions consist of the symbol for the element and the appropriate positive charge. The Lewis structures of monatomic negative ions consist of the symbol for the element with eight dots to represent the octet of valence electrons and the appropriate negative charge represented as a superscript.

Section 6.6

The formulas of ionic binary compounds are written by
determining the lowest common multiple of the positive and negative ionic charges and writing the symbols of each element followed by a subscript of the
number the charge of the respective ion must be multiplied by to give the common multiple of charge.
The symbol of the metal element is always written
first.

Section 6.7

Covalent bond   the chemical bond formed between two
atoms sharing the same pair of electrons. Compounds
with covalent bonds are much more numerous than
ionic compounds. Compounds formed between the atoms
of nonmetal elements are usually covalent compounds.

Section 6.8

Covalent compounds   compounds in which the atoms are
chemically bonded by a mutual attraction for the
same pair of electrons.

Molecule   a particle formed from atoms that are co-
valently bonded.

Section 6.9

Some of the properties of covalent compounds are:
  a. low melting points. Covalent compounds are often
     liquids or gases at normal temperatures. The
     melting points of solid covalent compounds are
     usually below 300°C.
  b. covalent compounds are nonconductors of both heat
     and electricity in the molten state.
  c. aqueous solution of covalent substances usually
     do not conduct electricity.

Section 6.10

Lewis structures of simple covalent compounds are drawn
using the symbols of the elements with dots to show
the bonding and nonbonding electrons. Frequently,

electron pairs are shown with a dash rather than two dots.

Section 6.11

The formulas of simple covalent binary compounds are determined by determining the lowest common multiple of covalent bonds formed by each of the atoms of the two elements and determining the ratio of the two atoms of each element from this common multiple.

Section 6.12

The electron repulsion theory is used to determine the geometry of simple covalent compounds.

A summary of electron repulsion and molecular geometry is shown below:

groups of electrons	geometry of the molecule	bond angles of the central atom
2	linear	180°
3	planar	120°
4	tetrahedral	109.5°
5	trigonal bipyrimidal	120° and 90°

Section 6.13

Electronegativity   the attraction an atom has for the electrons in a covalent bond. Fluorine atoms are the most electronegative of all the atoms of the elements. Electronegativity decreases with an increase of distance from fluorine in the Periodic Table.

Polar bond   a covalent bond between two atoms with different electronegativities which results in the uneven sharing of electrons leaving one of the atoms with a partial positive charge and the other atom with a partial negative charge.

Dipole   any particle that has a positive end or positive pole and a negative end or negative pole.

Dipolar or polar molecule   a molecule that has a positive side and a negative side.

Section 6.14

Polar molecules are the result of a combination of
   polar bonds and molecular geometry.

Section 6.15

Coordinate covalent bond   a covalent bond in which
   both bonding electrons are supplied by a single
   atom. Once the bond is formed a coordinate covalent
   bond is the same as any other covalent bond.

Section 6.16

Polyatomic ions   groups of atoms that are bonded by
   covalent bonds and which have either more or fewer
   electrons than the sum of electrons for the neutral
   atoms and therefore have an ionic charge.

Section 6.17

Formal charge   the unit charge assigned to an atom in
   a molecular structure. Formal charge = valence e^-
   of the atom - (1/2 shared e^- + unshared e^-).

Section 6.18

(See the problems for this section.)

Section 6.19

The charge on various ions can be determined from the
   charge on known ions they are associated with. Ions
   that you should know because of their position in
   the Periodic Table:
   a. The alkali metal or group IA elements, Li, Na, K,
      Rb, and Cs always form 1+ ions.
   b. The alkaline earth or group IIA elements, Be, Mg,
      Ca, Sr, Ba, and Ra always form 2+ ions.
   c. The group IIIA elements, Al, Ga, In, and Tl
      usually form 3+ ions.
   d. The halogens, F, Cl, Br, I, and At always form 1-
      ions.

e. Oxygen and sulfur form 2- ions. In a few cases oxygen will be encountered as a diatomic molecular O_2^{2-} ion, the peroxide ion.

SECTION 6.20

Metallic bonding   the bonding force which holds metal atoms in a solid structure.

Electron sea model of metals   the model of metallic bonding which has the kernels of metal atoms in regular geometric lattice positions of crystalline solid surrounded by a sea of the valence electrons.

SECTION 6.21

Nomenclature or names of compounds   the rules for naming simple inorganic compounds are listed in this section, text pages 214 - 217.

QUESTIONS AND PROBLEMS

SECTIONS 6.2 - 6.6   Ionic Bonding

1. Refer to Figure 6.3 on page 179 of the text and explain how the data in the graph verify the octet rule.

2. Using Lewis structures, show how ions would be formed by atoms of the following elements:
   a. In   b. Te   c. Br
   d. Cs   e. Ba

3. Indicate which ion in each of the following pairs is larger and explain the reason for your choice.
   a. Na^+ or Mg^{2+} b. Se^{2-} or Br^- c. N^{3-} or O^{2-}
   d. Mg^{2+} or Ca^{2+} e. Ca^{2+} or Zn^{2+} f. S^{2-} or Se^{2-}

4. Using the Periodic Table, give the Lewis structure for each ion and the formula for the compound formed from the following pairs of atoms:
   a. Ca and S   b. K and P   c. Ga and F
   d. In and O   e. Be and N

Answers:

1. In Figure 6.3, the data show a large increase in the

second ionization potential for $_{11}$Na, a large increase in the third ionization potential for $_{12}$Mg, a large increase in the third ionization potential for $_{13}$Al, a large increase in the fourth ionization potential for $_{14}$Si, and a large increase in the fifth ionization potential for $_{15}$P. In each of these cases, the large increase in ionization potential comes when it is necessary to break into the neon kernel which has an octet of electrons.

2. a. $\overset{..}{\text{In}}\cdot \ \rightarrow\ \text{In}^{3+} + 3\ e^-$ b. $\cdot \overset{..}{\text{Te}}\cdot\ +\ 2\ e^-\ \rightarrow\ :\overset{..}{\underset{..}{\text{Te}}}:^{2-}$

   c. $:\overset{..}{\underset{..}{\text{Br}}}.\ +\ 1\ e^-\ \rightarrow\ :\overset{..}{\underset{..}{\text{Br}}}:^{-}$ d. $\text{Cs}\cdot\ \rightarrow\ \text{Cs}^+\ +\ 1\ e^-$

   e. $\text{Ba}:\ \rightarrow\ \text{Ba}^{2+}\ +\ 2\ e^-$

3. a. Na^+ is larger because both ions have a neon kernel but Mg^{2+} has a larger positive charge.
   b. Se^{2-} is larger because both ions have a krypton configuration, but Se^{2-} has the larger negative charge.
   c. N^{3-} would be larger since both have a neon configuration, but N^{3-} has the larger negative charge.
   d. Ca^{2+} is larger since it has one more shell of electrons than Mg^{2+} and there is no difference in charge.
   e. Ca^{2+} is larger since the outer shell of both ions is the third shell and Zn^{2+} has a much larger nuclear charge.
   f. Se^{2-} is larger since it has an additional shell of electrons.

4. a. Ca^{2+} , $:\overset{..}{\underset{..}{\text{S}}}:^{2-}$, CaS   b. K^+ , $:\overset{..}{\underset{..}{\text{P}}}:^{3-}$, K_3P

   c. Ga^{3+} , $:\overset{..}{\underset{..}{\text{F}}}:^{-}$, GaF_3 d. In^{3+} , $:\overset{..}{\underset{..}{\text{O}}}:^{2-}$, In_2O_3

   e. Be^{2+} , $:\overset{..}{\underset{..}{\text{N}}}:^{3-}$, Be_3N_2

SECTIONS 6.7 - 6.11   Covalent Bonding

5. Draw the Lewis representations of the following covalent compounds:
   a. HI   b. I_2
   c. SiF_4 d. AsH_3
   e. CS_2 f. SeS_2

g. SiH_2Cl_2 h. $SeOCl_2$
i. Si_2H_6 j. BCl_3

6. Using the Periodic Table, write the formula for the
   covalent compound formed from the following pairs of
   elements.
   a. arsenic and oxygen
   b. silicon and sulfur
   c. selenium and hydrogen
   d. carbon and sulfur
   e. silicon and fluorine
   f. nitrogen and fluorine
   g. phosphorus and hydrogen
   h. carbon and selenium
   i. tellurium and bromine
   j. arsenic and selenium

7. For each of the following covalent compounds, give
   the shape of the molecule and the bond angle of the
   central atom.

   a. |O| b.   _ |O| _
        |O-Se-O| |Cl-Se-Cl|

   c.   _ |Cl| _   d. H-Se-H
        |Cl-Si-Cl|
          |Cl| e. Se=C=Se

8. For each of the following compounds, indicate the
   polar bonds and state whether the molecule is polar
   or nonpolar.

   a. |I-Cl| b.   H-S-H

   c.   _ |O| _   d. |Cl-N-Cl|
        |Cl-S-Cl| |Cl|

   e. |Cl-B-Cl| f. Se=C=Se
          |Cl|

9. Assign the formal charge to each atom in the following polyatomic ions and determine the net ionic
   charge on the ion.

   a. |S-H   b.   _ |O| _
                                    |O-S-O|

c.   H
     H-P-H
        H

d.   ‾
       ¦O¦   ‾
     ¦O-Cl-O¦
        ¦O¦
        ‾

Answers:

5. a. Hydrogen, H , forms one covalent bond; iodine,
       I , also forms one covalent bond.
                        H-I̅¦
                          ‾

    b. Each I can form one covalent bond.
                        ¦I̅-I̅¦
                        ‾ ‾

    c. Si will form four bonds; each F will form one
       bond.
                        ‾
                     _ ¦F¦ _
                     ¦F-S̲i-F¦
                       ¦F¦
                        ‾

    d. As will form three bonds; H will form one.
                        ‾
                     H-A̅s-H
                        H

    e. C will form four bonds and S will form two.
                     ‾   ‾
                     S̅=C=S̅

    f. Each sulfur requires two bonds so Se will expand
       its valence shell.
                     ‾   ‾
                     S̅=S̅e=S̅
                     ‾   ‾

    g. Si forms four bonds; each H and Cl forms one.
                     _   H   _
                    ¦Cl-S̲i-Cl¦
                     ‾   H   ‾

    h. Oxygen will form a double bond and each chlorine
       will form a single bond. Selenium will expand
       its valence shell to have ten electrons.
                     _ ¦O¦ _
                    ¦Cl-S̈e-Cl¦

    i. Each silicon will have four bonds and each hydro-
       ten one. The two silicons must be bonded to each
       other.

```
            H   H
            |   |
          H-Si-Si-H
            |   |
            H   H
```

j.
```
          |Cl-B-Cl|
            |Cl|
```

6. a. Arsenic needs three electrons for an octet; oxygen needs two. The lowest common multiple is six. Therefore, two arsenic atoms will bond with three oxygen atoms: As_2O_3

   b. Silicon needs four electrons for an octet; sulfur needs two. The lowest common multiple is four: SiS_2

   c. Selenium needs two electrons; hydrogen needs one. The lowest common multiple is two: H_2Se

   d. CS_2 e. SiF_4

   f. NF_3 g. PH_3

   h. CSe_2 i. $TeBr_2$

   j. As_2Se_3

7. a. Se has three groups of electrons; the molecule is planar and the bond angle is $120°$.

   b. Se has four sets of electrons; the molecule is tetrahedral and the bond angle is $109.5°$.

   c. Si has four groups of electrons; the molecule is tetrahedral and the bond angle is $109.5°$.

   d. Se has four groups of electrons; the molecule is tetrahedral and the bond angle is $109.5°$.

   e. C has two groups of electrons; the molecule is linear and the bond angle is $180°$.

8. a. |I-Cl| Chlorine is more electronegative and the molecule is polar.

   b.   H-S-H   The molecule is bent with a bond angle of about $109°$. The polar bonds do not cancel and the molecule is polar.

   c.   Cl-S-Cl   The molecule is tetrahedral and the polar bonds do not cancel; therefore, the molecule is polar.

d. $\overset{\longleftarrow}{} \overset{\longrightarrow}{}$

$|\overset{..}{\underset{..}{Cl}} - \overset{..}{\underset{..}{N}} - \overset{..}{\underset{..}{Cl}}|$
$|\overset{..}{\underset{..}{Cl}}|$ ↓

The molecule is not planar and the dipoles do not cancel; the molecule is polar.

e. $\overset{\longleftarrow}{} \overset{\longrightarrow}{}$

$|\overset{..}{\underset{..}{Cl}} - \overset{}{B} - \overset{..}{\underset{..}{Cl}}|$
$|\overset{..}{\underset{..}{Cl}}|$ ↓

The molecule is planar with bond angles of 120°. The polar bonds cancel each other. The molecule is nonpolar.

f. $\overset{\longleftarrow}{} \overset{\longrightarrow}{}$

$\underset{.}{Se} = C = \underset{.}{Se}$

The molecule is linear; the polar bonds cancel; and the molecule is nonpolar.

9. a. $\overset{1-}{}$

$|\overset{..}{\underset{.}{S}} - H$

Formal charge = valence e^- of the atom - (unshared e^- + 1/2 shared e^-)

For S: Formal charge = 6 - (6 + 1/2 x 2)

$= 6 - 7$

Formal charge = -1

b. S has a zero charge, the doubly bonded O has a zero charge, and the other two O's have a -1 charge. The charge on the ion is 2-.

c. P has a 1+ charge. H has no charge. The charge on the ion is 1+.

d. Each O is 1- and Cl is 3+. The net charge on the ion is 1-.

SECTION 6.18 Writing Formulas from Ionic Charge

10. Using the table on page 209 of the text and a Periodic Table, write the formulas of the following components:
   a. gallium sulfate
   b. barium phosphate
   c. indium nitrate
   d. ammonium sulfite
   e. sodium carbonate

SECTION 6.19 Determining Ionic Charge from Formulas

11. Give the charge of the polyatomic ion in the following compounds:
   a. K_2SeO_4
   b. $K_3C_6H_5O_7$
   c. $Mg_3(AsO_3)_2$
   d. $Ca(SCN)_2$

e. Ca_2SiO_4

Answers:

10. a. Ga^{3+} and SO_4^{2-} will form $Ga_2(SO_4)_3$

    b. Ba^{2+} and PO_4^{3-} will form $Ba_3(PO_4)_2$

    c. In^{3+} and NO_3^- will form $In(NO_3)_3$

    d. NH_4^+ and SO_3^{2-} will form $(NH_4)_2SO_3$

    e. Na^+ and CO_3^{2-} will form Na_2CO_3

11. a. Since there are two K^+, SeO_4 must be 2-.

    b. Since there are three K^+, $C_6H_5O_7$ must be 3-.

    c. Since there are three Mg^{2+}, each AsO_3 must be 3-.

    d. One Ca^{2+} indicates that each SCN is 1-.

    e. Two Ca^{2+} requires that SiO_4 is 4-.

SECTION 6.21   Naming Compounds

12. Give the formula of the following compounds:
    a. tin(II) nitrate
    b. tin (IV) nitrate
    c. ammonium phosphate
    d. sodium nitrite
    e. thallium(I) oxide
    f. rubidium cyanide
    g. calcium chloride
    h. lead(II) chloride
    i. barium hydroxide
    j. strontium fluoride
    k. sodium sulfide
    l. sodium sulfite
    m. sodium sulfate
    n. potassium chloride
    o. potassium hypochlorite
    p. potassium chlorite
    q. potassium chlorate
    r. potassium perchlorate

    s. rubidium carbonate
    t. potassium magnesium phosphate
    u. calcium hydrogen phosphate
    v. nitric acid
    w. hydroiodic acid
    x. boron trichloride
    y. dinitrogen pentasulfide
    z. nitrogen triiodide

13. Give the IUPAC name for the following compounds:

a. $NaHCO_3$	b. $NaBr$
c. K_2SO_4	d. K_3PO_4
e. Li_2CO_3	f. Li_2SO_3
g. CaS	h. BaI_2
i. BP	j. PF_5
k. H_2Te	l. TeO_3
m. H_3PO_4	n. $Ca(OCl)_2$
o. Tl_2O	p. Tl_2O_3
q. $GeCl_4$	r. GeO
s. $AuCl_3$	t. $AuBr$

Answers:

12.
a. $Sn(NO_3)_2$	b. $Sn(NO_3)_4$
c. $(NH_4)_3PO_4$	d. $NaNO_2$
e. Tl_2O	f. $RbCN$
g. $CaCl_2$	h. $PbCl_2$
i. $Ba(OH)_2$	j. SrF_2
k. Na_2S	l. Na_2SO_3
m. Na_2SO_4	n. KCl
o. $KClO$	p. $KClO_2$
q. $KClO_3$	r. $KClO_4$
s. Rb_2CO_3	t. $KMgPO_4$
u. $CaHPO_4$	v. HNO_3
w. HI	x. BCl_3
y. N_2S_5	z. NI_3

13. a. sodium hydrogen carbonate
    b. sodium bromide
    c. potassium sulfate
    d. potassium phosphate
    e. lithium carbonate
    f. lithium sulfite
    g. calcium sulfide
    h. barium iodide
    i. boron phosphide

      j. phosphorus pentafluoride
      k. hydrogen telluride
      l. tellurium trioxide
      m. phosphoric acid
      n. calcium hypochlorite
      o. thallium(I) oxide
      p. thallium(III) oxide
      q. germanium tetrachloride
      r. germanium oxide
      s. gold(III) chloride
      t. gold(I) bromide

SELF-TEST   Fill in the blanks:

1. Atoms or molecules with an excess or deficiency of electrons are called _____.
2. Negative ions are called _____.
3. Positive ions are called _____.
4. Alkaline earth or group IIA ions have charges of _____.
5. The melting point of ionic compounds is generally _____ than the melting point of covalent compounds.
6. The bond between two nonmetal atoms is usually a _____ bond.
7. The size of a potassium ion, K^+, is _____ than the size of a calcium ion, Ca^{2+}.
8. The covalent bond between two atoms with different electronegativities is a _____ bond.
9. The bond between two oppositely charged atomic particles is called a _____ bond.
10. Generally, the bond between a metal and a nonmetal is _____ bond.
11. The formal charge on the chlorine atom in the chlorate ion, $\mathrm{\overline{O}\text{-}\underset{\underset{\overline{\underline{O}}}{|}}{\underline{C}}l\text{-}\overline{O}}$, is a _____ charge.
12. The net charge on the chlorate ion is _____.
13. The formula for a binary compound of boron and phosphorus is _____.
14. The formula of a compound containing magnesium and nitrogen is _____.
15. The shape of a molecule in which the central atom is bonded to four other atoms by four groups of electrons is _____.
16. The bond angle of $\mathrm{\overline{F}\text{-}\underline{\overline{O}}\text{-}\overline{F}}$ is approximately _____.

17. The formula for copper(II) bromide is _____.
18. The formula of dinitrogen tetraoxide is _____.
19. The name of N_2O is _____.
20. The name of KH_2PO_4 is _____.

Answers:

1. ions
2. anions
3. cations
4. 2+
5. higher
6. covalent
7. smaller
8. polar
9. ionic
10. an ionic
11. 2+
12. 1-
13. BP
14. Mg_3N_2
15. tetrahedral
16. $109°$
17. $CuBr_2$
18. N_2O_4
19. dinitrogen oxide
20. potassium dihydrogen phosphate

7 Stoichiometry, the Quantitative Aspects of Chemical Reactions

OBJECTIVES
1. You should be able to calculate the formula or molecular weight of a compound (Section 7.1).
2. You should know that there are 6.022×10^{23} (Avogadro's number) of particles or entities in a mole of anything (Section 7.2).
3. You should be able to calculate the number of entities in multiples or fractions of a mole (Section 7.2).
4. You should be able to convert the quantity of a substance given in moles to grams (Section 7.3).
5. You should be able to convert the quantity of any substance given in grams to moles (Section 7.3).
6. You should be able to calculate the weight % composition of a substance from its formula (Section 7.4).
7. You should be able to calculate the empirical formula of a compound from its composition by weight (Section 7.5).
8. You should be able to calculate the molecular formula of a substance from its empirical formula and its molecular weight (Section 7.6).
9. You should be able to explain the meaning of the terms chemical reaction, reactants, and products (Section 7.7).
10. You should be able to write an equation for a chemical reaction from a verbal description of the reaction which includes the reactants and the products (Section 7.8).
11. You should be able to interpret an equation for a

chemical reaction (Section 7.8).

12. You should be able to balance simple chemical equations (Section 7.9).

13. You should be able to list the four general classes of simple inorganic reactions (Section 7.10).

14. You should be able to recognize each of these four general types of reactions from a chemical equation (Section 7.10).

15. You should be able to determine the mole ratio for any combination of reactants and products from a chemical reaction (Section 7.11).

16. You should be able to calculate any weight - weight relationship in a chemical reaction from the chemical equation for the reaction (Section 7.12).

17. You should be able to ascertain which reactant is present in the limiting quantity from the given weights of reactants and the chemical equation (Section 7.13).

18. You should be able to calculate the energy change associated with a reaction from the molar heats of reaction (Section 7.14).

IMPORTANT TERMS AND CONCEPTS

SECTION 7.1

Atomic weight   the average mass of all of the naturally occurring isotopes of an element relative to the mass of $^{12}_{6}C$. (review)

Formula weight   the sum of the atomic weights of all of the atoms that appear in the formula of an ionic substance which is the mass of the ions in a formula unit relative to the mass of $^{12}_{6}C$.

SECTION 7.2

Mole   that quantity of anything that has the same number of particles as the number of atoms in exactly 12 g of $^{12}_{6}C$ which is 6.022×10^{23} (Avogadro's number) particles. A mole is one of the basic units of measurement defined by the International System (SI) of Measurements.

Avogadro's number   the number of particles in a mole, 6.022×10^{23}. You should memorize Avogadro's number: 6.022×10^{23}.

Section 7.3

Gram <u>atomic</u> <u>weight</u>   the atomic weight of an element expressed in grams or the mass of 1 mole of atoms of a given element.

Gram <u>formula</u> <u>weight</u>   the formula weight of a compound expressed in grams or the mass of 1 mole of formula units.

Gram <u>molecular</u> <u>weight</u>   the molecular weight of a substance expressed in grams or the mass of 1 mole of molecules of any substance. Note: The gram atomic weight, the gram formula weight, and the gram molecular weight are the masses of moles of atoms, formula units, or molecules. This relationship is important because it permits us to determine the moles (numbers of particles) from the mass of the substance.

Section 7.4

(See the problems for this section.)

Section 7.5

<u>Empirical</u> <u>formula</u>   the simplest ratio of atoms of elements in a compound determined from experimental evidence.

Section 7.6

<u>Molecular</u> <u>formula</u>   the actual ratio of atoms in a molecule of a substance. The molecular formula is a whole number multiple of the empirical formula.

Section 7.7

<u>Chemical</u> <u>reaction</u>   the process by which one or more substances are transformed into one or more new substances.

<u>Reactants</u>   those substances that undergo the transformation in a chemical reaction.

<u>Products</u>   those substances that are formed as a result of a chemical reaction. All chemical reactions take place in a systematic manner in which bonds are made or broken in an orderly sequence.

SECTION 7.8

<u>Chemical equation</u>   a complete description of a chemi-
cal reaction using chemical formulas and symbols.

Common symbols:

+   means reacts with or is produced with
→   means to produce, yields, forms, etc.
↑   means is evolved as a gas
↓   means precipitates
Δ   means heat

Subscripts

(aq) aqueous or in water solution
(g)   gas
(1)   liquid
(s)   solid

SECTION 7.9

Rules for balancing simple equations by inspection:
a. Start with the <u>correct</u> <u>formulas</u> for all of the
<u>reactants</u> and <u>products</u>.
b. Inspect both sides of the equation to ascertain
which elements are not balanced.
c. Balance the metal elements first.
d. Balance the nonmetal elements with the exception
of oxygen and hydrogen.
e. Polyatomic ions which appear unchanged in the
reaction are balanced as a unit.
f. Balance the hydrogens.
g. If the oxygens are not balanced by now, check all
other elements to make sure they are balanced be-
fore balancing oxygen separately.
h. When it is necessary to change the number of
atoms of a previously balanced element, the number of atoms of that element must be changed on
both sides of the equation.

SECTION 7.10

<u>Combination reaction</u>   a reaction in which two reac-
tants combine to form a single product.
<u>Decomposition reaction</u>   a chemical reaction in which a

single substance is transformed into two or more
    simpler substances.

<u>Single</u> <u>displacement</u> <u>reaction</u>   a simple substance re-
    acts with a compound to displace or substitute for
    another simple substance.

<u>Double</u> <u>displacement</u> <u>(metathesis)</u>   a reaction which in-
    volves the exchange of ions of one compound for sim-
    ilarly charged ions of another compound with one of
    the two products being either water, a gas, or an
    insoluble solid.

Section 7.11

The coefficients of the reactants and products in a
    balanced chemical equation give the molar ratios of
    reactants and products.

Section 7.12

(See problems for this section.)

Section 7.13

<u>Limiting</u> <u>reagent</u>   the reactant that is present in the
    lesser amount according to the required molar ratio.
    The quantities of products formed must be calculated
    from the quantity of the limiting reagent.

Section 7.14

<u>Exothermic</u> <u>reaction</u>   a reaction which is accompanied
    by the release of heat.

<u>Endothermic</u> <u>reaction</u>   a reaction which takes heat from
    its surroundings.

<u>Enthalpy</u> <u>of</u> <u>reaction</u> (ΔH)   the change in heat content
    of the reaction system. The change in enthalpy (ΔH)
    is positive for an endothermic reaction and negative
    for an exothermic reaction.

<u>Joule</u> (J)   the unit of energy in the International
    System (SI).

<u>Calorie</u>   the unit of heat in the Metric system defined
    as that quantity of heat that will raise the temper-
    ature of 1 g of water from $14.5^{\circ}C$ to $15.5^{\circ}C$. 1 ca-
    lorie = 4.184 J.

QUESTIONS
AND
PROBLEMS

SECTION 7.1 Formula and Molecular Weights

1. Determine the molecular or formula weight of the following substances:
   a. KCl   b. C_6H_6
   c. $Ba_3(PO_4)_2$ d. $CuSO_4 \cdot 10\ H_2O$

SECTION 7.2 The Mole

2. Calculate the following:
   a. the number of atoms in 1×10^{-6} moles of C.
   b. the number of moles of H atoms in 2 moles of table sugar, $C_{12}H_{22}O_{11}$.
   c. the number of atoms of H in 2 moles of table sugar, $C_{12}H_{22}O_{11}$.

SECTION 7.3 Measuring the Mole- Gram Atomic Weight, Gram-Molecular Weight, Gram-Formula Weight

3. Calculate the number of moles of substance in each of the following quantities:
   a. 69.0 g Na   b. 1 000 g H_2O
   c. 490 g H_2SO_4 d. 30.10 g $Ba_3(PO_4)_2$

4. Calculate the mass of the following quantities of substances:
   a. 55.5 mole H_2O b. 0.500 mole $C_{12}H_{22}O_{11}$
   c. 0.600 mole Na_3PO_4 d. 0.750 mile Mg_2N_3

Answers:

1. For molecular or formula weights, the whole is the sum of all its parts. Molecular or formula weights are equal to the sum of the atomic weights for each atom in the unit.

   a.   K = 39.10 amu
   _____ Cl = 35.45 amu_
   mol. wt.   = 74.55 amu

   b. C_6H_6
      6 x C = 6 x 12.01 = 72.06 amu
      6 x H = 6 x 1.01 = 6.06 amu
                mol. wt. = 78.12 amu

c. $Ba_3(PO_4)_2$

$$
\begin{aligned}
3 \times Ba &= 3 \times 137.3 = 411.9 \text{ amu} \\
2 \times P &= 2 \times 31.0 = 62.0 \text{ amu} \\
8 \times O &= 8 \times 16.0 = 128.0 \text{ amu} \\
\hline
&\qquad \text{mol. wt.} = 601.9 \text{ amu}
\end{aligned}
$$

d. $CuSO_4 \cdot 10\ H_2O$

$$
\begin{aligned}
1 \times Cu &= 1 \times 63.5 = 63.5 \text{ amu} \\
1 \times S &= 1 \times 32.0 = 32.0 \text{ amu} \\
4 \times O &= 4 \times 16.0 = 64.0 \text{ amu} \\
20 \times H &= 20 \times 1.0 = 20.0 \text{ amu} \\
10 \times O &= 10 \times 16.0 = 160.0 \text{ amu} \\
\hline
&\qquad \text{mol. wt.} = 339.5 \text{ amu}
\end{aligned}
$$

2. a. 1 mole atoms = 6.02×10^{23} atoms

1×10^{-6} ~~mole atoms~~ $\times \dfrac{6.02 \times 10^{23} \text{ atoms}}{1 \text{ ~~mole atoms~~}}$

$= 6.02 \times 10^{17}$ atoms

b. 2 mole ~~$C_{12}H_{22}O_{11}$~~ $\times \dfrac{22 \text{ H}}{1 \text{ ~~$C_{12}H_{22}O_{11}$~~}} = 44$ mole H atoms

c. 2 ~~mole~~ $C_{12}H_{22}O_{11} \times \dfrac{6.02 \times 10^{23} \text{ ~~molecules~~}}{1 \text{ ~~mole~~} C_{12}H_{22}O_{11}}$

$\times \dfrac{22 \text{ H atoms}}{1 \text{ ~~molecule~~}} = 2.65 \times 10^{25}$ H atoms

3. a. 1 mole Na = 23 g

69 ~~g Na~~ $\times \dfrac{1 \text{ mole Na}}{23 \text{ ~~g Na~~}} = 3.0$ mole Na

b. 1 mole H_2O = 18.02 g H_2O

1 000 ~~g H_2O~~ $\times \dfrac{1 \text{ mole } H_2O}{18 \text{ ~~g H_2O~~}} = 55.49$ mole H_2O

c. 1 mole H_2SO_4 = 98.0 g H_2SO_4

490 ~~g H_2SO_4~~ $\times \dfrac{1 \text{ mole } H_2SO_4}{98.0 \text{ ~~g H_2SO_4~~}} = 5.00$ mole H_2SO_4

d. 0.05001 mole $Ba_3(PO_4)_2$

4. a. 55.5 ~~mole H_2O~~ $\times \dfrac{18.0 \text{ g } H_2O}{1 \text{ ~~mole H_2O~~}} = 999$ g H_2O

b. 1 mole $C_{12}H_{22}O_{11}$ = 342 g $C_{12}H_{22}O_{11}$

$$0.500 \; \cancel{mole \; C_{12}H_{22}O_{11}} \; \times \; \frac{342 \text{ g } C_{12}H_{22}O_{11}}{1 \; \cancel{mole \; C_{12}H_{22}O_{11}}}$$

$$= 171 \text{ g } C_{12}H_{22}O_{11}$$

c. 98.4 g Na_3PO_4

d. 68.0 Mg_2N_3

SECTION 7.4   Percentage Composition from Chemical
Formulas

5. Calculate the percentage composition of each element
in each of the following compounds:
   a. Na_2SO_3 b. $Al_2(SO_4)$
   c. $KMgPO_4$ d. $K_2Cr_2O_7$

SECTION 7.5   Determination of the Empirical Formula of
a Compound

6. From the weight percent composition of the following
compounds calculate the empirical formula of the
compound:
   a. 40.0% S and 60% O
   b. 25.9% N and 74.1% O
   c. 42.1% Na , 18.9% P , and 39.0% O
   d. 60.4% Rb , 22.7% S , and 17.0% O

7. A 7.200 g sample of a compound containing only os-
mium and oxygen was found to contain 5.388 g of Os.
What is the empirical formula of the compound?

8. A 1.623 g sample of a compound containing only tita-
nium and chlorine was found to contain 0.420 g Ti.
What is the empirical formula of the compound?

SECTION 7.6   The Determination of the Molecular Formula
from the Empirical Formula and the Molecu-
lar Weight

9. A compound consisting of 50.7% C , 4.25% H , and
45.1% O was found to have an approximate molecular
weight of 140.
   a. What is the empirical formula of the compound?
   b. What is the molecular formula of the compound?
   c. What is the exact molecular weight of the

compound?

10. A compound consisting of 49.0% C , 2.72% H , and
    48.3% Cl was found to have a molecular weight of
    approximately 145.
    a. What is the empirical formula of the compound?
    b. What is the molecular formula of the compound?
    c. What is the exact molecular weight of the com-
       pound?

Answers:

5. a. Na_2SO_3

   Calculate the molecular weight of the compound:
   2 x Na = 2 x 23.0 = 46.0
   1 x S = 1 x 32.0 = 32.0
   3 x O = 3 x 16.0 = 48.0
                mol. wt. =126.0

   The wt.% of each element is the ratio of the mass
   of the atoms of the element in the compound to
   the molecular weight of the compound x 100%.

   % Na = $\frac{46.0}{126.0}$ x 100% = 36.5% Na

   % S = $\frac{32.0}{126.0}$ x 100% = 25.4% S

   % O = $\frac{48.0}{126.0}$ x 100% = $\underline{38.1\% \text{ O}}$
                                        100.0%

   Note: Always check your calculations to see that
   all of the percentages add to 100%.

   b. $Al_2(SO_4)_3$ 15.8% Al , 28.1% S , 56.2% O

   c. $KMgPO_4$ 24.7% K , 15.5% Mg , 19.6% P ,
                        40.5% O

   d. $K_2Cr_2O_7$ 26.6% K , 35.4% Cr , 38.1% O

6. a. If we assume 100 g of compound, we have 40.0 g S
      and 60.0 g O.
      First, we find the molar ratio of atoms:

      40.0 g S x $\frac{1 \text{ mole S}}{32.0 \text{ g S}}$ = 1.25 mole S

      60.0 g O x $\frac{1 \text{ mole O}}{16.0 \text{ g O}}$ = 3.75 mole O

We convert the molar ratios to whole numbers by dividing each number by the smaller of the two:

$$\frac{1.25 \text{ mole S}}{1.25} = 1 \text{ mole S}$$

$$\frac{3.75 \text{ mole O}}{1.25} = 3 \text{ mole O}$$

The empirical formula is SO_3

b. $25.9 \text{ g N} \times \dfrac{1 \text{ mole N}}{14.0 \text{ g N}} = 1.85 \text{ mole N}$

$74.1 \text{ g O} \times \dfrac{1 \text{ mole O}}{16.0 \text{ g O}} = 4.63 \text{ mole O}$

Convert the ratios to whole numbers:

$$\frac{1.85 \text{ mole N}}{1.85} = 1 \text{ mole N} \times 2 = 2 \text{ mole N}$$

$$\frac{4.63 \text{ mole O}}{1.85} = 2.5 \text{ mole O} \times 2 = 5 \text{ mole O}$$

The empirical formula is N_2O_5

In this case, when the ratio came out to be a fraction after we divided each of the ratios by the smaller, it was necessary to double the ratios to get a whole number. All of the ratios were doubled. The fractions that are most common and which you should be alert for are:

0.5 and 1.5 which are doubled
0.33 and 0.67 which are tripled
0.25 and 0.75 which are multiplied by 4

c. Na_3PO_4

d. $Rb_2S_2O_3$

7. Given: wt. Os = 5.388 g
           wt. compound = 7.200 g

First find the weight of oxygen in the compound:
    7.200 g compound - 5.388 g Os = 1.812 g O

Find the empirical formula from the weights of the elements:

$5.388 \text{ g Os} \times \dfrac{1 \text{ mole Os}}{190.2 \text{ g Os}} = 0.0283 \text{ mole Os}$

$$1.812 \; \cancel{g \; O} \times \frac{1 \text{ mole O}}{16.0 \; \cancel{g \; O}} = 0.113 \text{ mole O}$$

Convert the molar ratios to whole numbers by dividing the ratios by the smaller number:

$$\frac{0.0283 \text{ mole Os}}{0.0283} = 1 \text{ mole Os}$$

$$\frac{0.113 \text{ mole O}}{0.0283} = 4 \text{ mole O}$$

The empirical formula is OsO_4

8. $TiCl_4$

9. a. Find the empirical formula of the compound from the % composition.
   The empirical formula is $C_3H_3O_2$

   b. To determine the molecular formula:
   Find the formula weight of the empirical formula.

   $C_3H_3O_2$ formula weight = 71

   Find the multiple the empirical formula must be multiplied by so that it will give the approximate molecular weight.

   $140 = 71 \times n$
   $\quad n = 2$

   $(C_3H_3O_2)_n = C_6H_6O_4$

   c. Calculate the exact molecular weight from the molecular formula and the atomic weights of the elements.

   exact molecular weight = 142.12

10. a. C_3H_2Cl b. $C_6H_4Cl_2$
    c. exact molecular wt. = 147.0

SECTIONS 7.8 AND 7.9   Chemical Equations

11. Write chemical equations for each of the following reactions:
    a. Sodium hydrogen sulfite, $NaHSO_{3(s)}$, decomposes to form sodium sulfite, $Na_2SO_{3(s)}$; sulfur dioxide, SO_2 ; and water.

b. Ammonia, NH_3 , reacts with magnesium to produce magnesium nitride, Mg_3N_2 , and hydrogen gas.

c. Ammonia, $NH_{3(g)}$, reduces copper(II) oxide, $CuO_{(s)}$, to form nitrogen, copper, and water.

d. Silane, SiH_4 , burns in the oxygen of the air to form silicon dioxide, SiO_2 , and water.

e. Diboron trioxide reacts with magnesium to give boron and magnesium oxide.

12. Balance the following equations:

a. $Al_2O_3 + F_2 \rightarrow AlF_3 + OF_2$

b. $Ca_3(PO_4)_2 + C \rightarrow Ca_3P_2 + CO$

c. $Al_{(s)} + H_2SO_{4(aq)} \rightarrow Al_2(SO_4)_{3(aq)} + H_{2(g)}$

SECTION 7.10   Types of Reactions

13. Classify each of the following reactions and balance the equation:

a. $SO_{2(g)} + O_{2(g)} \rightarrow SO_{3(g)}$

b. $NH_4NO_{2(s)} \rightarrow N_{2(g)} + H_2O_{(g)}$

c. $H_2O_{(g)} + C_{(s)} \rightarrow CO_{(g)} + H_{2(g)}$

d. $NH_4Br_{(aq)} + AgNO_{3(aq)} \rightarrow AgBr_{(s)} + NH_4NO_{3(aq)}$

e. $CaCO_{3(s)} \rightarrow CaO_{(s)} + CO_{2(g)}$

f. $KOH + CO_2 \rightarrow KHCO_3$

g. $H_2S_{(g)} + CuCl_{2(aq)} \rightarrow CuS_{(s)} + HCl_{(aq)}$

h. $H_2O_{(1)} + Ca_{(s)} \rightarrow Ca(OH)_{2(aq)} + H_{2(g)}$

i. $(NH_4)_2CO_{3(aq)} + CaCl_{2(aq)} \rightarrow NH_4Cl_{(aq)} + CaCO_{3(s)}$

j. $Fe_3O_{4(1)} + C_{(s)} \rightarrow Fe_{(1)} + CO_{(g)}$

Answers:

11. a. $2 \text{ NaHSO}_{3(s)} \rightarrow \text{Na}_2\text{SO}_{3(s)} + \text{SO}_{2(g)} + \text{H}_2\text{O}_{(g)}$

    b. $2 \text{ NH}_{3(g)} + 3 \text{ Mg}_{(s)} \rightarrow \text{Mg}_3\text{N}_{2(s)} + 3 \text{ H}_{2(g)}$

    c. $2 \text{ NH}_{3(g)} + 3 \text{ CuO}_{(s)} \rightarrow 3 \text{ Cu}_{(s)} + \text{N}_{2(g)} + 3 \text{ H}_2\text{O}_{(g)}$

    d. $\text{SiH}_{4(g)} + 2 \text{ O}_{2(g)} \rightarrow \text{SiO}_{2(s)} + 2 \text{ H}_2\text{O}_{(g)}$

    e.

12. a. $\text{Al}_2\text{O}_3 + 6 \text{ F}_2 \rightarrow 2 \text{ AlF}_3 + 3 \text{ OF}_2$

    b. $\text{Ca}_3(\text{PO}_4)_2 + 8 \text{ C} \rightarrow \text{Ca}_3\text{P}_2 + 8 \text{ CO}$

    c. $2 \text{ Al}_{(s)} + 3 \text{ H}_2\text{SO}_{4(aq)} \rightarrow \text{Al}_2(\text{SO}_4)_{3(aq)} + 3 \text{ H}_{2(g)}$

13. a. $\text{SO}_{2(g)} + \text{O}_{2(g)} \rightarrow \text{SO}_{3(g)}$ combination

    b. $\text{NH}_4\text{NO}_{2(s)} \rightarrow \text{N}_{2(g)} + 2 \text{ H}_2\text{O}_{(g)}$ decomposition

    c. $\text{H}_2\text{O}_{(g)} + \text{C}_{(s)} \rightarrow \text{CO}_{(g)} + \text{H}_{2(g)}$ single displace-
ment

    d. $\text{NH}_4\text{Br}_{(aq)} + \text{AgNO}_{3(aq)} \rightarrow \text{AgBr}_{(s)} + \text{NH}_4\text{NO}_{3(aq)}$
                           double displacement

    e. $\text{CaCO}_{3(s)} \rightarrow \text{CaO}_{(s)} + \text{CO}_{2(g)}$ decomposition

    f. $\text{KOH} + \text{CO}_2 \rightarrow \text{KHCO}_3$ combination

    g. $\text{H}_2\text{S}_{(g)} + \text{CuCl}_{2(aq)} \rightarrow \text{CuS}_{(s)} + 2 \text{ HCl}_{(aq)}$ double
                            displacement

    h. $2 \text{ H}_2\text{O}_{(l)} + \text{Ca}_{(s)} \rightarrow \text{Ca(OH)}_{2(aq)} + \text{H}_{2(g)}$ single
                            displacement

    i. $(\text{NH}_4)_2\text{CO}_{3(aq)} + \text{CaCl}_{2(aq)} \rightarrow 2 \text{ NH}_4\text{Cl}_{(aq)} + \text{CaCO}_{3(s)}$
                            double displacement

    j. $\text{Fe}_3\text{O}_{4(l)} + 4 \text{ C}_{(s)} \rightarrow 3 \text{ Fe}_{(l)} + 4 \text{ CO}_{(g)}$ single
                            displacement

SECTION 7.11   Mole Ratios from the Coefficients of a
                    Balanced Chemical Equation

14. Give the mole ratio of H_3PO_3 to each of the two

products for the reaction:

$$4 H_3PO_3 \rightarrow 3 H_3PO_4 + PH_3$$

15. Give the mole ratio of iron to each of the two re-
    actants in the reaction:

$$Fe_3O_{4(l)} + 4 C_{(s)} \rightarrow 3 Fe_{(l)} + 4 CO_{(g)}$$

16. In the reaction:

$$Al_2O_3 + 6 F_2 \rightarrow 2 AlF_3 + 3 OF_2$$

    a. How many moles of F_2 will be required to react
       with 1.5 mole Al_2O_3?
    b. How many moles of AlF_3 will be produced?
    c. How many moles of OF_2 will be produced?

17. In the reaction:

$$3 BaCl_2 + 2 Na_3PO_4 \rightarrow Ba_3(PO_4)_2 + 6 NaCl$$

    a. How many moles of Na_3PO_4 will be required to re-
       act with 0.6 moles of $BaCl_2$?
    b. How many moles of $Ba_3(PO_4)_2$ will be produced from
       0.6 moles of $BaCl_2$?
    c. How many moles of NaCl will be produced?

SECTION 7.12   Weight - Weight Calculations in Chemical
                Reactions

18. What weight of copper(II) sulfide, CuS , will be
    formed when 0.2118 g of copper is burned with an ex-
    cess of sulfur?

19. What weight of potassium chlorate, $KClO_3$, must be
    decomposed to KCl and O_2 to produce 1.600 g of O_2?

20. What weight of calcium oxide is obtained when
    1.00 kg of calcium carbonate is decomposed to cal-
    cium oxide and carbon dioxide?

21. How many grams of calcium carbide, CaC_2 , will be
    required to produce 0.520 g of acetylene, C_2H_2?

SECTION 7.13 Limiting Reagents

22. What weight of carbon dioxide will be produced when a solution containing 0.730 g of HCl is added to 1.26 g of sodium hydrogen carbonate, $NaHCO_3$?

$$HCl_{(aq)} + NaHCO_{3(s)} \rightarrow NaCl_{(aq)} + H_2O + CO_{2(g)}$$

23. a. What weight of acetylene, C_2H_2 , will be produced when 1.802 g of water are added to 16.02 g of calcium carbide, CaC_2 ?

   b. What weight of acetylene will be produced if an additional 9.010 g of water are added to the remaining calcium carbide?

SECTION 7.14 Heats of Reaction

24. What quantity of heat is produced when 500 g of acetylene, C_2H_2 , is burned if the heat of combustion for 1 mole of acetylene is 312 kcal?

Answers:

14. From the balanced equation:

$$\frac{4 \text{ mole } H_3PO_3}{3 \text{ mole } H_3PO_4} \quad \text{and} \quad \frac{4 \text{ mole } H_3PO_3}{1 \text{ mole } PH_3}$$

15. From the balanced equation:

$$\frac{3 \text{ mole Fe}}{1 \text{ mole } Fe_3O_4} \quad \text{and} \quad \frac{3 \text{ mole Fe}}{4 \text{ mole C}}$$

16. a. $1.5 \text{ mole } Al_2O_3 \times \dfrac{6 \text{ mole } F_2}{1 \text{ mole } Al_2O_3} = 9.0 \text{ mole } F_2$

   b. $1.5 \text{ mole } Al_2O_3 \times \dfrac{2 \text{ mole } AlF_3}{1 \text{ mole } Al_2O_3} = 3.0 \text{ mole } AlF_3$

   c. $1.5 \text{ mole } Al_2O_3 \times \dfrac{3 \text{ mole } OF_2}{1 \text{ mole } Al_2O_3} = 4.5 \text{ mole } OF_2$

17. a. 0.4 mole Na_3PO_4

   b. 0.2 mole $Ba_3(PO_4)_2$

c. 1.2 mole NaCl

18. a. Write a balanced equation for the reaction:

$$Cu + S \rightarrow CuS$$

b. Convert the weight of the known substance to moles:

$$0.218 \; \cancel{g\;Cu} \times \frac{1 \; mole \; Cu}{63.54 \; \cancel{g \; Cu}} = 0.00333 \; mole \; Cu$$

c. Determine the mole ratio of the unknown substance to the known substance:

$$\frac{1 \; mole \; CuS}{1 \; mole \; Cu}$$

d. From the mole ratio $\frac{1 \; mole \; CuS}{1 \; mole \; Cu}$ and the moles of

Cu (0.00333 mole) find the moles of unknown:

$$\frac{1 \; mole \; CuS}{1 \; \cancel{mole \; Cu}} \times 0.00333 \; \cancel{mole \; Cu} = 0.00333 \; mole \; CuS$$

e. Convert moles of CuS to weight in grams:

$$0.00333 \; mole \; CuS \times \frac{95.55 \; g \; CuS}{1 \; mole \; CuS} = 0.3182 \; g \; CuS$$

Note: We can combine the procedures:

$$0.218 \; \cancel{g\;Cu} \times \frac{1 \; \cancel{mole \; Cu}}{63.54 \; \cancel{g \; Cu}} \times \frac{1 \; \cancel{mole \; CuS}}{1 \; \cancel{mole \; Cu}} \times \frac{95.55 \; g \; CuS}{1 \; \cancel{mole \; CuS}}$$

<u>convert known quantity</u> to moles   <u>multiply by</u> the mole ratio of the unknown to the known   <u>convert the</u> moles of unknown to weight in g

= 0.3182 g CuS

19. $2 \; KClO_3 \rightarrow KCl + 3 \; O_2$

$$1.600 \; \cancel{g \; O_2} \times \frac{1 \; \cancel{mole \; O_2}}{32.0 \; \cancel{g \; O_2}} \times \frac{2 \; \cancel{mole \; KClO_3}}{3 \; \cancel{mole \; O_2}} \times \frac{122.6 \; g \; KClO_3}{1 \; \cancel{mole \; KClO_3}}$$

<u>convert known quantity</u> to moles   <u>multiply by</u> the mole ratio of the unknown to the known   <u>convert the</u> moles of unknown to weight in g

= 4.083 g KClO₃

20. 560 g CaO

21. 1.283 g CaC$_2$

22. a. Calculate the moles of each of the reactants:

0.730 ~~g HCl~~ $\times \dfrac{1 \text{ mole HCl}}{36.5 \text{ ~~g HCl~~}} = 0.02$ mole HCl

1.26 ~~g NaHCO$_3$~~ $\times \dfrac{1 \text{ mole NaHCO3}}{84.0 \text{ ~~g NaHCO$_3$~~}} = 0.015$ mole NaHCO$_3$

b. From the mole ratio in the balanced equation, determine the limiting reactant:

0.02 ~~mole HCl~~ $\times \dfrac{1 \text{ mole NaHCO3}}{1 \text{ ~~mole HCl~~}} = 0.02$ mole NaHCO$_3$

0.02 mole HCl would require 0.02 mole NaHCO$_3$. Since there are only 0.015 mole HaHCO$_3$, the limiting reactant is NaHCO$_3$.

c. Using the limiting reactant, calculate the weight of the product:

0.015 ~~mole NaHCO$_3$~~ $\times \dfrac{1 \text{ ~~mole CO$_2$~~}}{1 \text{ ~~mole NaHCO3~~}} \times \dfrac{44.0 \text{ g CO}_2}{\text{~~mole CO$_2$~~}}$

$= 0.660$ g CO$_2$

23. a. Calculate the limiting reactant:

1.802 ~~g H$_2$O~~ $\times \dfrac{1 \text{ mole H}_2\text{O}}{18.02 \text{ ~~g H$_2$O~~}} = 0.1000$ mole H$_2$O

16.02 ~~g CaC$_2$~~ $\times \dfrac{1 \text{ mole CaC}_2}{64.10 \text{ ~~g CaC$_2$~~}} = 0.2500$ mole CaC$_2$

0.1 ~~mole H$_2$O~~ $\times \dfrac{1 \text{ mole CaC}_2}{2 \text{ ~~mole H$_2$O~~}} = 0.05000$ mole CaC$_2$

H$_2$O is the limiting reagent.

0.1000 ~~mole H$_2$O~~ $\times \dfrac{1 \text{ ~~mole C$_2$H$_2$~~}}{2 \text{ ~~mole H$_2$O~~}} \times \dfrac{26.04 \text{ g C}_2\text{H}_2}{1 \text{ ~~mole C$_2$H$_2$~~}}$

$= 1.301$ g C$_2$H$_2$

b. The remaining CaC$_2$ is:

0.2500 mole CaC$_2$ - 0.0500 mole CaC$_2$ = 0.2000 CaC$_2$

$$9.010 \text{ g } \cancel{H_2O} \times \frac{1 \text{ mole } H_2O}{18.02 \text{ } \cancel{g H_2O}} = 0.5000 \text{ mole } H_2O$$

$$0.5000 \text{ } \cancel{\text{mole } H_2O} \times \frac{1 \text{ mole } CaC_2}{2 \text{ } \cancel{\text{mole } H_2O}} = 0.2500 \text{ mole } CaC_2$$

CaC_2 is the limiting reagent for the second addition of water:

$$0.2000 \text{ } \cancel{\text{mole } CaC_2} \times \frac{1 \text{ } \cancel{\text{mole } C_2H_2}}{1 \text{ } \cancel{\text{mole } CaC_2}} \times \frac{26.04 \text{ g } C_2H_2}{1 \text{ } \cancel{\text{mole } C_2H_2}}$$

$$= 5.208 \text{ g } C_2H_2$$

24. $500 \text{ } \cancel{\text{g } C_2H_2} \times \frac{1 \text{ } \cancel{\text{mole } C_2H_2}}{26.0 \text{ } \cancel{\text{g } C_2H_2}} \times \frac{312 \text{ kcal}}{1 \text{ } \cancel{\text{mole } C_2H_2}} = 6000 \text{ kcal}$

SELF-TEST

1. What is the molecular weight of $Co_2(SO_4)_3$?
2. How many hydrogen atoms are there in 0.020 mole of $C_6H_{12}O_6$?
3. How many moles of mercury(II) iodate, $Hg(IO_3)_2$, are there in 0.188 g?
4. What is the percentage nickel in podomite, Ni_3S_4 ?
5. A compound contains 43.4% K , 20.0% C , 1.1 % H , and 35.5% O . What is the empirical formula of the compound?
6. What is the weight of 0.125 mole of $Ca_3(PO_4)_2$?
7. A compound has an empirical formula of C_2HBr and an approximate molecular weight of 300. What is its molecular formula?

The equation: $C_3H_8O + O_2 \rightarrow CO_2 + H_2O$ pertains to Questions 8 - 11.

8. What is the coefficient for H_2O in the balanced equation?
9. What is the mole ratio of O_2 to C_3H_8O in the balanced equation?
10. What weight of oxygen is required to react with 1.20 g of C_3H_8O ?
11. What weight of water will be produced when 1.20 g of C_3H_8O react with excess O_2 ?
12. The reaction: $3 Fe + 2 O_2 \rightarrow Fe_3O_4$ is what type of reaction?
13. The reaction: $Br_2 + H_2S \rightarrow 2 HBr + S$ is what type of reaction?
14. The reaction: $NH_4NO_2 \rightarrow N_2 + 2 H_2O$ is what type of

reaction?

The following problem pertains to Questions 15 - 17.

Hydrogen gas is commonly generated from HCl and zinc according to the reaction: $Zn + 2 HCl \rightarrow ZnCl_2 + H_2$. When 25 grams of zinc are added to a solution containing 50 grams of HCl:

15. What substance is the limiting reagent?
16. What weight of $ZnCl_2$ will be produced?
17. How many grams of HCl or Zn will remain after the reaction is complete?
18. If the heat of combustion of acetylene, C_2H_2, is 312 kcal/mole, what weight of acetylene must be burned to obtain 156 kcal of heat?

The following problem pertains to Questions 19 - 20.

When 1.217 g of antimony combine with sulfur the resulting compound weighs 1.697 g.

19. What weight of sulfur is in the compound?
20. What is the empirical formula of the compound?

Answers:

1. 405.9
2. 1.44×10^{23}
3. 2.5×10^{-4} mole
4. 57.91%
5. $K_2C_3H_2O_4$
6. 38.8 g
7. $C_6H_3Br_3$
8. 8
9. $\dfrac{9 \text{ mole } O_2}{2 \text{ mole } C_3H_8O}$
10. 2.88 g O_2
11. 1.44 g H_2O
12. combination
13. single displacement
14. decomposition
15. Zn
16. 52.1 g $ZnCl_2$
17. 23.2 g HCl
18. 13.0 g
19. 0.480 g S
20. Sb_2S_3

chapter

8 The Behavior of Gases

OBJECTIVES
1. You should be able to give the description of the gaseous state according to the Kinetic-Molecular Model (Section 8 1).
2. You should be able to relate the qualitative behavior of gases to the Kinetic-Molecular Model of Gases (Section 8.2).
3. You should be able to describe the atmosphere (Section 8.3).
4. You should be able to state the meaning and give the value of a standard atmosphere (1 atmosphere = 760 torr = 1.013×10^5 pascal) (Section 8.3).
5. You should be able to state the four parameters of a gas (Section 8.4).
6. You should be able to describe the various methods of measuring pressure (Section 8.5).
7. You should be able to relate the pressure of a gas to its volume at constant temperature; i.e., know and be able to use Boyle's Law (Section 8.6).
8. You should be able to relate the volume of a gas at constant pressure to its absolute temperature; i.e., know and be able to apply Charles' Law (Section 8.7).
9. You should know and be able to explain the meaning of the absolute temperature (Section 8.7).
10. You should be able to relate the pressure of a given volume of gas to its absolute temperature (Section 8.8).
11. You should be able to relate the pressure, temperature, and volume changes for a given quantity of

gas; i.e., know and be able to apply the combined gas laws (Section 8.9).

12. You should be able to explain the Law of Combining Volumes (Section 8.10).

13. You should know Avogadro's Hypothesis (Section 8.11).

14. You should be able to explain what is meant by an ideal gas (Section 8.12).

15. You should know how each of the four parameters of a gas is related through the Ideal Gas Law (Section 8.12).

16. You should be able to solve any problem using the Ideal Gas Law (Section 8.12).

17. You should know exactly what is meant by Standard Conditions or Standard Temperature and Pressure (Section 8.13).

18. You should know the value of the molar volume of a gas at STP (Section 8.13).

19. You should be able to calculate the molecular weight of a gas from its vapor density (Section 8.14).

20. You should be able to explain Dalton's Law of Partial Pressures using the Kinetic-Molecular Model (Section 8.16).

21. You should be able to relate the mass and the average velocity of molecules of two gases (Graham's Law of Effusion) (Section 8.17).

IMPORTANT TERMS AND CONCEPTS

SECTION 8.1

The <u>Kinetic-Molecular Theory</u>   the theory that explains the states of matter as they are related to the particulate nature of matter and the molecular motion caused by the heat content of molecules. (See Section 8.1 of the text for a complete statement of this theory.)

SECTION 8.3

<u>Atmosphere</u>   the envelope of gases that surround the Earth.

<u>Atmospheric pressure</u>   the pressure exerted in all directions by the gases of the atmosphere as a result of the gravitational attraction of the Earth.

<u>Standard atmosphere</u>   the average atmospheric pressure which is equal to 1 atmosphere = 760 torr = 1.013×10^5 pascal.

Atmosphere   a common unit of pressure of a gas.

Torr   a common unit of pressure for gases which is equal to the pressure exerted by a column of mercury exactly 1 mm in height.

Pascal   the SI unit for pressure (1 torr = 1.333×10^2 pascal).

SECTION 8.4

Parameters of a gas   the four variables that define the states of a gas: quantity (mole), volume (liter), pressure (torr, atmosphere, or pascal), and absolute temperature (kelvin).

Torricellian barometer   a device for measuring atmospheric pressure consisting of a long glass tube filled with mercury and inverted with the open end of the tube in a dish of mercury, also called a mercury barometer. (See Figure 8.1 of the text.)

Manometer   a device for measuring pressure. (See Figure 8.2 of the text.)

SECTION 8.6

Boyle's Law   for a given quantity of gas at a constant temperature, the volume is proportional to the inverse of the pressure. The mathematical statements of Boyle's Law are as follows:

$$V \propto \frac{1}{P} \qquad \text{or} \qquad V = k \times \frac{1}{P} \qquad \text{or} \qquad PV = k$$

Boyle's Law can be used to relate the change in volume brought about by a change in pressure at a constant temperature for a fixed quantity of gas:

$$P_1 V_1 = P_2 V_2 \qquad \text{where } P_1 \text{ and } V_1 \text{ are the initial pressure and volume and } P_2 \text{ and } V_2 \text{ are the final pressure and volume.}$$

SECTION 8.7

Charles' Law   for a given quantity of gas at constant pressure, the volume is proportional to the absolute temperature. The mathematical statements of Charles' Law are as follows:

$$V \propto T \quad \text{or} \quad V = kT \quad \text{or} \quad \frac{V}{T} = k$$

Charles' Law is most frequently applied to relate the volume and temperature of a fixed quantity of gas at constant pressure at two sets of conditions:

$$\frac{V_1}{T_1} = \frac{V_2}{T_2}$$

Absolute temperature the temperature scale that uses absolute zero as its zero point. (The SI temperature unit, the kelvin, is on an absolute scale.)

Absolute zero the temperature at which, according to the Kinetic-Molecular Theory, all molecular motion stops.

Section 8.8

Pressure-temperature dependence at a constant volume, the pressure of a fixed quantity of gas is proportional to the absolute temperature ($P \propto T$). From this relationship, we can derive the following equation for the variation of pressure with temperature for a fixed quantity of a gas at constant volume:

$$\frac{P_1}{T_1} = \frac{P_2}{T_2}$$

Section 8.9

Combined gas law the combination of gas laws pertaining to a fixed quantity of gas into a single mathematical expression:

$$\frac{P_1 V_1}{T_1} = \frac{P_2 V_2}{T_2}$$

Section 8.10

Gay-Lussac's Law of Combining Volumes when two gases combine chemically at constant temperature and pressure, they always combine in a fixed ratio of small whole numbers by their volumes.

Section 8.11

<u>Avogadro's</u> <u>hypothesis</u>   at any fixed temperature and pressure, equal volumes of gases contain an equal number of particles.

Section 8.12

<u>Ideal</u> <u>Gas</u> <u>Law</u>   a combination of all of the gas laws incorporating the fourth variable, the quantity of gas: $PV = nRT$.

In the equation,

$$PV = nRT$$

P is the pressure of the gas;
V is the volume of the gas;
n is the quantity of gas, i.e., the number of particles;
T is the absolute temperature;
R is the proportionality constant, called the universal ideal constant. The numerical value of R is dependent on the units used for P, V, n, and T. (See Table 8.3 of the text.)

Section 8.13

<u>Standard</u> <u>conditions</u> <u>or</u> <u>standard</u> <u>temperature</u> <u>and</u> <u>pressure</u> <u>(STP)</u>   273 K (0°C) and 1 atmosphere (760 torr or 1.013×10^5 pascal).

<u>Molar</u> <u>volume</u>   the volume of 1 mole of an ideal gas at STP or 22.4 liters.

Section 8.14

<u>Vapor</u> <u>density</u>   the density of a substance in the gaseous state at a specified temperature and pressure.

Section 8.15

(See problems for this section.)

Section 8.16

<u>Dalton's</u> <u>Law</u> <u>of</u> <u>Partial</u> <u>Pressures</u>   the total pressure exerted by a mixture of gases is the sum of the

partial pressures exerted by each gas in the mixture.

SECTION 8.19

Graham's <u>Law</u> <u>of</u> <u>Effusion</u>   the ratio of the average velocities of the molecules of two gases at the same temperature is proportional to the inverse of the ratio of the square roots of their respective masses.

$$\frac{v_1}{v_2} = \frac{\sqrt{m_2}}{\sqrt{m_1}}$$

QUESTIONS
AND
PROBLEMS

SECTION 8.1   The Kinetic-Molecular Model of Gases

1. List the postulates of the Kinetic-Molecular Theory.

SECTION 8.2   The Qualitative Behavior of Gases

2. Explain the following facts by the Kinetic-Molecular Theory:
   a. a gas expands to occupy any volume of a container;
   b. a gas is easily compressed;
   c. the density of gases is low;
   d. the volume of a fixed quantity of gas at a constant pressure expands when heated.

SECTION 8.4   The Parameters of a Gas

3. List the four variable measurements that define the behavior of a gas.

SECTION 8.6   Volume-Pressure Relationships — Boyle's Law

4. A sample of O_2 gas with a volume of 2.5 liters and a pressure of 760 torr is compressed to a volume of 1.0 liter. What is the new pressure of the gas if there is no change in temperature?

5. A balloon filled with helium has a volume of 3.0 liters at a pressure of 760 torr at $20^{\circ}C$. What is the volume of the balloon at a pressure of 253 torr at $20^{\circ}C$?

SECTION 8.7 Volume-Temperature Relationships —
Charles' Law

6. A balloon filled with air has a volume of 50 liters
at 75°F and 1 atmosphere. What is the volume of the
balloon at -5°F and 1 atmosphere?

7. If a gas with a volume of 14 liters at 20°C is al-
lowed to expand at a constant pressure, what will
its volume be at 100°C?

SECTION 8.8 Pressure-Temperature Relationships

8. If the pressure in a tire on a summer day is 46
lb/in.² at 70°F, assuming that the volume of the
tire does not change, what will the tire pressure be
after a 1 hour drive when the temperature of the
tire has increased to 158°F?

9. If the pressure in a sealed can is 1.0 atmosphere at
20°C, what will the pressure inside the can be if it
is thrown into a fire and heated to 500°C?

SECTION 8.9 Volume-Temperature-Pressure Relationships—
The Combined Gas Laws

10. A passenger car tire has a volume of 22 liters when
the pressure of the air inside the tire is 3.2 atmospheres and the temperature of the tire is 20°C.
Calculate the pressure inside the tire after it has
been driven for 1 hour if the temperature of the
tire is 70°C and the volume has expanded to 23
liters.

11. What will the volume of an expansible balloon be at
an altitude of 15 000 meters where the temperature
is -60°C and the pressure of the gas inside the balloon is 150 torr, if the original volume of the balloon was 3.20 m³ at 27°C and an original pressure of
750 torr?

Answers:

1. See Section 8.1 in the text.

2. a. The molecules of a gas are in rapid straight-line

motion and continue so until they collide with
another molecule or the walls of the container.
The attractive forces between the molecules of a
gas are negligible in comparison to their kinetic
energy. Therefore, the molecules of a gas will
continue to expand until they are restricted by
the walls of the container.

b. The actual volume of molecules of a gas is insignificant compared to the total volume of the gas.
   Since the molecules of a gas are so widely separated, the gas is easily compressed.

c. The actual volume of the molecules of a gas is
   insignificant compared to the total volume of the
   gas. Therefore, a gas is mostly open space with
   a low density.

d. Since the average kinetic energy of the gas molecules is proportional to the absolute temperature
   of the gas, as the temperature increases more
   molecules hit the walls of the expandible container with greater force causing an increase in
   volume.

3. Quantity (moles), volume (liters or m^3), temperature
   (K), and pressure (atmospheres, torr, or pascal).

4. Note: Problems involving the gas laws or a change
   in one or more of the parameters of a gas affecting
   another of the parameters may be solved either by
   the rote use of memorized equations of the gas laws
   or by the rationale of analyzing how the change will
   affect the given parameter and developing a correction factor accordingly. In any case, you should
   always check your answer to make sure it is reason-
   able: i.e., if the pressure increases, the volume
   decreases; if the pressure decreases, the volume
   increases.

   Given: P_1 = 760 torr V_1 = 2.5 liters
                 P_2 = ? V_2 = 1.0 liter

   By rote: $P_1V_1 = P_2V_2$
                 760 torr x 2.5 liters = P_2 x 1.0 liter
                 P_2 = 1900 torr

By analyzing the change and correcting for it with a correction factor:

$$P_2 = 760 \text{ torr} \times \frac{2.5 \text{ } \cancel{\text{liters}}}{1.0 \text{ } \cancel{\text{liter}}}$$

Note: The volume decreases; the pressure must increase. The correction factor is greater than 1.

5. v_2 = 9.0 liters. The pressure decreased; the volume must increase.

6. In any case where we use temperature with gas problems, the temperature must be the absolute temperature in kelvins (K).

Convert $^{\circ}$F to K

$$(f - 32)\frac{5}{9} = c \qquad f \text{ is } ^{\circ}F \text{ and } c \text{ is } ^{\circ}C$$

$$^{\circ}C + 273 = K$$

$$75\,^{\circ}F = 24\,^{\circ}C = 297 \text{ K}$$

$$-5\,^{\circ}F = -21\,^{\circ}C = 252 \text{ K}$$

Given: v_1 = 50 liters v_2 = ?
        T_1 = 297 K T_2 = 252 K

$$\frac{v_1}{T_1} = \frac{v_2}{T_2}$$

$$v_2 = v_1 \times \frac{T_2}{T_1}$$

$$v_2 = 50 \text{ liters} \times \underbrace{\frac{252 \text{ K}}{297 \text{ K}}}_{\substack{\text{correction for} \\ \text{temperature change}}}$$

$$v_2 = 42 \text{ liters}$$

The temperature decreases; the volume must decrease.

7. 17.8 liters. The temperature increased; the volume must increase.

8. Convert the temperatures to K

$$70^\circ F = 21^\circ C = 294 \text{ K}$$
$$158^\circ F = 70^\circ C = 343 \text{ K}$$

Given: $P_1 = 46$ lb/in.^{2} $T_1 = 294$ K
        $P_2 = ?$ $T_2 = 343$ K

$$\frac{P_1}{T_1} = \frac{P_2}{T_2}$$

$$P_2 = 46 \text{ lb/in.}^2 \times \frac{343 \text{ K}}{294 \text{ K}}$$

$$P_2 = 54 \text{ lb/in.}^2$$

The temperature increased; therefore, the pressure must increase.

9. 2.83 atmospheres. The temperature increased; the pressure must increase.

10. Convert the temperature to K

$$70^\circ C = 343 \text{ K}$$
$$20^\circ C = 293 \text{ K}$$

Given: $V_1 = 22$ liters $P_1 = 3.20$ atmospheres
        $T_1 = 293$ K
        $V_2 = 23$ liters $P_2 = ?$
        $T_2 = 343$ K

$$\frac{P_1 V_1}{T_1} = \frac{P_2 V_2}{T_2}$$

Solve for P_2:

$$P_2 = P_1 \times \frac{V_1}{V_2} \times \frac{T_2}{T_1}$$

Alternately: $P_2 = P_1 \times$ Volume correction $\times$ T correction

$$P_2 = 3.20 \text{ atmospheres} \times \frac{22 \text{ \sout{liters}}}{23 \text{ \sout{liters}}} \times \frac{343 \text{ K}}{273 \text{ K}}$$

$$\begin{bmatrix} \text{volume} \\ \text{increases;} \\ \text{correction} \\ \text{factor} \\ < 1 \end{bmatrix} \begin{bmatrix} \text{temperature} \\ \text{increases;} \\ \text{correction} \\ \text{factor} \\ > 1 \end{bmatrix}$$

$$P_2 = 3.6 \text{ atmospheres}$$

11. $$V_2 = 3.20 \text{ m}^3 \times \frac{750 \text{ \sout{torr}}}{150 \text{ \sout{torr}}} \times \frac{213 \text{ K}}{300 \text{ K}}$$

$$\begin{bmatrix} \text{pressure} \\ \text{decreases;} \\ \text{correction} \\ > 1 \end{bmatrix} \begin{bmatrix} \text{temperature} \\ \text{decreases;} \\ \text{correction} \\ < 1 \end{bmatrix}$$

$$V_2 = 11.4 \text{ m}^3$$

SECTION 8.12 - 8.15 Applications of the Ideal Gas Law

12. What is the weight of H_2 gas in a cylinder that has a volume of 25 liters and a pressure of 270 atmospheres at 25°C?

13. What is the volume of a balloon filled with 20 g of helium if the pressure is 150 torr and the temperature is -50°C?

14. What is the molar concentration (moles/liter) of hydrogen at 450°C and a pressure of 300 atmospheres?

15. If air has an average molecular weight of 29, what is the difference in weight of a volume of 50.0 liters of air at 20°C and 740 torr and the same volume of helium at the same temperature and pressure?

16. What is the molecular weight of a gas if a 2.42 g sample of its vapor occupies 530 ml at 745 torr and 350 K?

17. Calculate the molecular weight of a substance if 1.89 g of its vapor has a volume of 255 ml at 758 torr and 372 K.

18. What volume of O_2 at 25°C and 1 atmosphere is required to burn a liter of gasoline (C_8H_{18}) which has a density of 0.870 g/ml if the only products of the reaction are CO_2 and H_2O? What volume of CO_2 is produced?

Answers:

12. Since three of the four required parameters of a gas are given, this problem can be solved using the ideal gas law.

Given: P = 270 atm V = 25 liters
         n = ? T = 298 K
         R = 0.0820 $\dfrac{\text{liter atm}}{\text{mole K}}$

$PV = nRT$

270 ~~atm~~ x 25 ~~liters~~ = n x 0.082 $\dfrac{\text{liter atm}}{\text{mole }\cancel{\text{K}}}$ x 298 ~~K~~

n = 276 mole

276 mole H_2 x $\dfrac{2 \text{ g } H_2}{1 \text{ mole } H_2}$ = 552 g H_2

13.   Given: P = 150 torr V = ?
                n = 5 mole T = 223 K
                R = 62.4 $\dfrac{\text{liter torr}}{\text{mole K}}$

$PV = nRT$

150 ~~torr~~ x V = 5 ~~mole~~ x 62.4 $\dfrac{\text{liter } \cancel{\text{torr}}}{\cancel{\text{mole K}}}$ x 223 ~~K~~

V = 464 liters

14.   Given: P = 300 atm V = 1 liter
                n = ? T = 623 K
                R = 0.082 $\dfrac{\text{liter atm}}{\text{mole K}}$

$PV = nRT$

300 ~~atm~~ x 1 ~~liter~~ = n x 0.082 $\dfrac{\text{liter atm}}{\text{mole }\cancel{\text{K}}}$ x 623 ~~K~~

n = 0.17 mole

molar concentration = $\dfrac{0.17 \text{ mole}}{1 \text{ liter}}$ = 0.17 mole/liter

15. Plan: Calculate the moles of gas. From the moles
    of gas calculate the weights of air and helium. De-
    termine the difference in weight.

    Given: P = 740 torr V = 50.0 liter
            n = ? T = 293 K
            R = 62.4 $\dfrac{\text{liter torr}}{\text{mole K}}$

    $PV = nRT$

    740 ~~torr~~ x 50.0 ~~liter~~ = n x 62.4 $\dfrac{\text{liter torr}}{\text{mole K}}$ x 293 K̶

    n = 2.02 mole

    Air: 2.02 ~~mole~~ x $\dfrac{29\ g}{1\ \text{mole}}$ = 58.7 g

    He: 2.02 ~~mole~~ x $\dfrac{4\ g}{1\ \text{mole}}$ = 8.1 g
                                               _____

                Difference = 50.6 g

16. Plan: Find the number of moles. Determine the
    molecular weight from the given weight for the cal-
    culated number of moles.

    Given: P = 745 torr V = 0.530 liter
            n = ? T = 350 K
            R = 62.4 $\dfrac{\text{liter torr}}{\text{mole K}}$

    $PV = nRT$

    745 ~~torr~~ x 0.530 ~~liter~~ = n x 62.4 $\dfrac{\text{liter torr}}{\text{mole K}}$ x 350 K̶

    n = 0.0181 mole

    $\dfrac{2.42\ g}{0.0181\ \text{mole}}$ = 134 g/mole = molecular weight

17. 227 g/mole

18. Plan: Calculate the number of moles of O_2 from the
    known weight of gasoline and a balanced equation for
    the reaction. Find the volume of O_2 using the ideal
    gas law.

        2 C_8H_{18} + 25 O_2 → 16 CO_2 + 18 H_2O

Calculate the number of moles of O_2:

$$1 \; \cancel{\text{liter } C_8H_{18}} \times \frac{1\,000 \; \cancel{\text{ml}}}{1 \; \cancel{\text{liter}}} \times \frac{0.870 \; \cancel{\text{g } C_8H_{18}}}{1 \; \text{ml } \cancel{C_8H_{18}}}$$

$$\times \frac{1 \; \cancel{\text{mole } C_8H_{18}}}{114 \; \cancel{\text{g } C_8H_{18}}} \times \frac{25 \; \text{mole } O_2}{2 \; \cancel{\text{mole } C_8H_{18}}} = 728 \; \text{mole } O_2$$

Using the Ideal gas law:

Given: $P = 1$ atm $V = ?$
   $n = 728$ mole O_2 $T = 298$ K
   $R = 0.082 \; \dfrac{\text{liter atm}}{\text{mole K}}$

$PV = nRT$

$$1 \; \cancel{\text{atm}} \times V = 728 \; \cancel{\text{mole}} \times 0.082 \; \frac{\text{liter } \cancel{\text{atm}}}{\cancel{\text{mole K}}} \times 298 \; \cancel{K}$$

$V = 17\,800$ liter O_2

11 400 liters of CO_2 are produced

Section 8.16

19. A 10.0 liter container is filled with 3 moles of H_2
    and 1 mole of N_2 at $450^{\circ}C$:
    a. What is the pressure of each gas?
    b. What is the total pressure in the container?
    c. What is the pressure of NH_3 if 1/2 of the N_2 and
       H_2 react to form NH_3 ?
    d. What is the total pressure in the container after
       the reaction?

Section 8.17   Graham's Law

20. If oxygen molecules have an average velocity of 1084
    mi/h at $25^{\circ}C$, what is the average velocity of hydro-
    gen molecules at the same temperature?

21. If the average velocity of helium atoms at $25^{\circ}C$ is
    1360 m/s at $25^{\circ}C$, what is the average velocity of
    methane, CH_4, molecules at the same temperature?

Answers:

19. Plan: The pressure of each gas can be calculated

using the ideal gas law. According to Dalton's law of partial pressures, the total pressure is the sum of the partial pressures of each of the gases.

a. For H_2: P = ? v = 10.0 liter
                 n = 3.00 moles T = 723 K
                 R = 0.082 $\dfrac{\text{liter atm}}{\text{mole K}}$

$PV = nRT$

P x 10.0 ~~liter~~ = 3 ~~mole~~ x 0.082 $\dfrac{\text{liter atm}}{\text{\sout{mole K}}}$ x 723 K

P_{H_2} = 17.8 atm

For N_2: P_{N_2} = 5.9 atm

b. P_{total} = P_{H_2} + P_{N_2} = 17.8 atm + 5.9 atm = 23.7 atm

c. From the balanced equation for the reaction we find that 1.5 mole H_2 and 0.5 mole N_2 produce 1 mole NH_3.

P_{NH_3} = 8.9 atm

d. After the reaction the container has 1.5 mole H_2, 0.5 mole N_2, and 1 mole NH_3.

P_{H_2} = 8.9 atm P_{N_2} = 2.9 atm P_{NH_3} = 5.9 atm

P_{total} = 17.7 atm

20.   Given: v_{O_2} = 1084 mi/h m_{H_2} = 2.0 m_{O_2} = 32.0

$$\frac{v_{H_2}}{v_{O_2}} = \frac{\sqrt{m_{O_2}}}{\sqrt{m_{H_2}}}$$

$$\frac{v_{H_2}}{1084 \text{ mi/h}} = \frac{\sqrt{32}}{\sqrt{2}}$$

v_{H_2} = 4336 mi/h

21. v_{CH_4} = 481 m/s

SELF-TEST   1. As the absolute temperature of a gas increases the average kinetic energy of its molecules _____
2. At a constant temperature, the volume of a fixed quantity of gas _____ when the pressure decreases.
3. At a constant pressure the volume of a fixed quantity of gas _____ when the temperature decreases.
4. When a sealed container of a gas is heated the pressure inside the container _____.
5. At the same temperature, molecules of CH_4 have a _____ velocity than molecules of O_2.
6. If the volume of an ideal gas is 2.5 liters at 1 atmosphere of pressure, what is the pressure of the gas when the volume increases to 10.0 liters?
7. If the volume of a gas is 0.5 liter at 760 torr, what is the volume of the same gas at 700 torr?
8. If the quantity and the volume of a gas are fixed and the pressure of the gas is 1520 torr at $327^{\circ}C$, what will the pressure of the gas be at $27^{\circ}C$?
9. A balloon has a volume of 30.0 liters at $27^{\circ}C$; what will its volume be at the same pressure if it is cooled to $-23^{\circ}C$?

The following statement pertains to Questions 10 - 13:
   If 100 g of $N_{2(g)}$ and 100 g of $O_{2(g)}$ are placed in separate containers of equal volume at the same temperature, make the following comparisons of the two gas samples: M.W. N_2 = 28, O_2 = 32.

10. Compare the number of molecules in each container.
11. Compare the pressure in each of the two flasks.
12. Compare the average speed of the molecules in the two flasks.
13. Compare the average kinetic energy of the molecules of the two gases.
14. If a perfectly expansible balloon has a volume of 10.0 m^3 at 760 torr and $27^{\circ}C$, what will the volume of the balloon be when the pressure is 380 torr and the temperature is $-23^{\circ}C$?
15. What is the volume of 1 mole of gas at STP?
16. What is the density of CO_2 gas in g/liter at STP?
17. Nitrogen and hydrogen combine to form ammonia according to the following equation:

$$N_{2(g)} + 3\ H_{2(g)} \rightarrow 2\ NH_{3(g)}$$

What volume of H_2 is required to react with 5 liters of N_2 at 100 atm and $400^\circ C$?

18. How many liters of NH_3 would be produced from the N_2 and H_2 in Question 17?

19. What volume will a gas sample occupy at 700 torr and $27^\circ C$ if there is 0.020 mole of gas in the sample?
    $(R = 62.4 \frac{\text{liter torr}}{\text{mole K}})$

20. What is the molecular weight of a gas if 2.50 g of the gas has a volume of 0.500 liter and a pressure of 750 torr at 350 K? $(R = 62.4 \frac{\text{liter torr}}{\text{mole K}})$

Answers:

1. increases
2. increases
3. decreases
4. increases
5. greater
6. 0.25 atm
7. 5.4 liter
8. 760 torr
9. 25 liters
10. more molecules of N_2
11. pressure is greater for N_2
12. the average speed of the N_2 molecules is greater
13. the average kinetic energy of the molecules is the same
14. 16.7 liter
15. 22.4 liter
16. 1.96 g/liter
17. 15 liters H_2
18. 10 liters NH_3
19. 0.535 liters
20. 145

9 Liquids and Solids

OBJECTIVES
1. You should be able to compare the differences in the liquid state and the gaseous state according to the Kinetic-Molecular Theory (Section 9.1).
2. You should be able to explain the properties of the liquid state according to the Kinetic-Molecular Theory (Section 9.1).
3. You should be able to describe London dispersion forces (Section 9.2).
4. You should know the factors that cause increases in the London dispersion forces (Section 9.2).
5. You should be able to describe polar attractions between molecules (Section 9.3).
6. You should be able to make qualitative predictions of the relative polarity of molecules (Section 9.3).
7. You should be able to describe what is meant by hydrogen bonding between molecules (Section 9.4).
8. You should be able to recognize those substances that have hydrogen bonding between molecules (Section 9.4).
9. You should be able to explain evaporation according to the Kinetic-Molecular Theory for liquids (Section 9.5).
10. You should be able to describe equilibrium vapor pressure according to the Kinetic-Molecular Theory (Section 9.5).
11. You should be able to relate the equilibrium vapor pressure of a substance to the attractive forces between its molecules (Section 9.6).

12. You should be able to relate the equilibrium vapor pressure of a liquid to Dalton's theory of partial pressures (Section 9.6).
13. You should be able to calculate equilibrium vapor pressures from volume changes, Dalton's law of partial pressure, and the gas laws (Section 9.6).
14. You should be able to define the normal boiling point of a liquid (Section 9.7).
15. You should be able to compare the normal boiling points of different substances based on the variation of one of the intermolecular attractive forces (Section 9.7).
16. You should be able to relate the properties of the solid state to the Kinetic-Molecular Theory of solids (Section 9.8).
17. You should be able to describe ionic solids and give examples (Section 9.9).
18. You should be able to describe molecular solids and give examples (Section 9.10).
19. You should be able to describe metallic solids and give examples (Section 9.11).
20. You should be able to describe network solids and give examples (Section 9.12).
21. You should be able to explain the vapor pressure of a solid according to the Kinetic-Molecular Theory (Section 9.13).
22. You should be able to relate the process of melting or freezing to the Kinetic-Molecular Theory of solids (Section 9.14).
23. You should be able to give operational definitions of specific heat, heat of fusion, and heat of vaporization (Section 9.15).
24. You should be able to give qualitative descriptions of the transition from the solid state to the liquid state and the gaseous state (Section 9.15).
25. You should be able to explain on a qualitative basis the energy changes for any transition between physical states (Section 9.15).
26. You should be able to calculate the quantitative energy changes for any transition between physical states of matter.

IMPORTANT
TERMS
AND
CONCEPTS

SECTION 9.1

Liquid state   the state of matter that is characterized by the following properties:

    a. the lack of a definite shape
    b. ease of displacement by other matter
    c. a fixed volume
    d. the ability to flow in response to a force such
       as gravity.

<u>Kinetic-Molecular</u> <u>Theory</u> <u>of</u> <u>liquids</u>   the theory that
explains the states of matter according to its par-
ticulate nature and the molecular motion caused by
heat content. The postulates of the Kinetic-Molecular Theory that apply to all three states of matter
and the postulates that apply only to the liquid
state are listed in Section 9.1 of the text. You
should go over them carefully until you understand
them.

SECTION 9.2

<u>London</u> <u>dispersion</u> <u>forces</u>   the intermolecular forces
that result from the momentary uneven distribution
of electrons within the molecules. London dispersion forces increase with molecular size and weight.

SECTION 9.3

<u>Polar</u> <u>attraction</u> <u>or</u> <u>dipole-dipole</u> <u>forces</u>   the inter-
molecular attractive forces caused by the electro-
static attraction of polar molecules. Review Sections 6.13 and 6.14 in the text on Polar Bonds and
Polar and Nonpolar Molecules.

SECTION 9.4

<u>Hydrogen</u> <u>bonding</u>   the strong attractive force between
a hydrogen atom bonded to either a nitrogen, oxygen,
or fluorine atom and another nitrogen, oxygen, or
fluorine atom.

SECTION 9.5

<u>Evaporation</u>   the escape of high-energy molecules from
the liquid phase into the gaseous phase. Cooling
accompanies the evaporation of a liquid because the
average kinetic energy of the molecules of the liquid decreases as the higher-energy molecules escape.

Section 9.6

Dynamic equilibrium   the condition in which two
exactly equal but opposite changes take place at the
same rate causing what appears to be a static or un-
changing condition.

Equilibrium vapor pressure   the partial pressure of
the gas phase that is in equilibrium with the liquid
phase at a given temperature. Note: The term
"vapor pressure of a liquid" is frequently used in-
stead of the term equilibrium vapor pressure.

Volatile liquid   liquids that have high vapor pressures at normal temperatures and therefore evaporate
rapidly.

Section 9.7

Boiling   the process that involves the transition from
the liquid state to the vapor state accompanied by
the turbulent bubbling in the liquid phase.

Boiling point   the temperature at which the vapor
pressure of a liquid is equal to the pressure of the
gas phase in contact with it.

Normal boiling point   the boiling point of a liquid at
an external pressure of 1 atmosphere. The greater
the intermolecular attractive forces between the
molecules of a substance, the higher the boiling
point.

Section 9.8

Solid   the physical state of matter characterized by
the following properties:
a. rigid shape
b. fixed volume
c. restriction of other substances from the space
occupied
d. incompressibility

Modifications of the Kinetic-Molecular Theory for
Solids
a. the average kinetic energy of the particles of a
solid are low compared to the attractive forces
between particles;
b. the particles of a solid are held in crystal lattice positions by strong attractive forces

between particles;

c. the particles of a solid no longer have unrestricted motion;

d. the motion of particles is restricted to vibrational and rotational motion.

Section 9.9

Ionic <u>solid</u>   a solid in which the lattice particles are oppositely charged ions and the attractive force between particles is the ionic bond. The properties of ionic solids are (a) high melting point; (b) hard, brittle crystals. Examples: $NaCl$, $CuSO_4$, NH_4Cl , $CaCl_2$, etc.

Section 9.10

Molecular <u>solid</u>   a solid in which the lattice particles are molecules and the attractive force of the lattice are the common intermolecular attractive forces. The properties of molecular solids are (a) moderately low melting points; (b) soft crystals. Examples: ice, table sugar, aspirin, moth crystals, etc.

Section 9.11

Metallic <u>solid</u>   solids in which the particles are the atomic kernels and the lattice forces are the electrostatic attraction of electrons in an electron sea. The properties of metallic solids are (a) high thermal conductivity; (b) malleability and ductility. Examples: metallic elements and alloys.

Section 9.12

Network <u>or</u> covalent <u>solids</u>   solids in which the lattice particles are atoms and the attractive forces are covalent bonds. The properties of network or covalent solids are (a) extremely high melting points; (b) extremely hard crystals. Examples: diamonds, C ; carborundum, SiC ; quartz, SiO_2 ; etc. Note: See Table 9.1 in the text for a summary of crystalline solids.

SECTION 9.13

Vapor pressure of a solid   the partial pressure of the gas phase of the substance that is in equilibrium with the solid state. The vapor pressure of most solids is extremely low.

Sublimation   the transition of a substance from the solid state directly to the vapor state.

SECTION 9.14

Melting   the transition from the solid state to the liquid state.

Freezing   the transition from the liquid state to the solid state.

Melting or freezing point   the temperature at which there is an equilibrium between the solid state of the substance and the liquid state. The melting point of a solid is a function of the strength of the attractive forces in the structure of the crystal lattice. Melting points are relatively independent of the external pressure on a substance.

SECTION 9.15

Heat capacity   the heat required to increase the temperature of 1 g of a substance $1^{\circ}C$ or 1 K.

Specific heat   the ratio of the heat capacity of a substance to the heat capacity of water at $15^{\circ}C$.

Heat of fusion   the heat required to convert a given quantity of substance from the solid state to the liquid state.

Molar heat of fusion   the heat required to convert 1 mole of a substance from the solid state to the liquid state.

Molar heat of vaporization   the heat required to convert 1 mole of a substance from the liquid state into the vapor state.

QUESTIONS
AND
PROBLEMS

SECTION 9.2 - 9.4   Intermolecular Attractive Forces

1. Give the attractive forces between molecules of each of the following substances:

a. H_3C-O-H
b. $SiCl_4$
c. Ne
d. H_2S
e. $H-O-O-H$
f. H_2CCl_2

2. For each of the following pairs of substances, indicate which substance has the greater attractive force between its molecules and explain why:
a. Ne and Xe
b. H_3C-O-H and H_3CF
c. CF_4 and H_2CCl_2

SECTION 9.5   Evaporation

3. Explain why the temperature of a volatile liquid such as ether decreases when it is poured from a storage bottle into an evaporating dish.

SECTION 9.6   Vapor Pressure

4. When enough of a small amount of a liquid is injected into a 1 liter closed system at $20^{\circ}C$ so that about 0.5 ml of liquid remains, the pressure in the system increases from 755 torr to 832 torr. What is the equilibrium vapor pressure of the liquid at $20^{\circ}C$?

5. How many moles of H_2 collected over water at $25^{\circ}C$ and 759 torr are contained in a 500 ml flask? (The vapor pressure of water at $25^{\circ}C$ is 24 torr.)

6. If a small amount of liquid is injected into a system at a constant pressure of 760 torr with an expansible volume at $20^{\circ}C$, the volume increases from 250 ml to 276 ml. What is the vapor pressure of the liquid if a few drops of liquid remain?

SECTION 9.7   The Normal Boiling Point

7. Predict the order of the boiling points for the following groups of substances:

a. $SiBr_4$, $SiCl_4$, SiF_4 , SiH_4 , SiI_4

b. H_3C-CH_3 , H_3CF , H_3C-O-H

Answers:

1. a. H_3C-O-H has London dispersion forces, dipolar forces, and hydrogen bonding.
   b. $SiCl_4$ has only London dispersion forces; the molecule has polar bonds but because of its geometry is nonpolar.
   c. Ne has only London dispersion forces.
   d. H_2S has London dispersion forces and dipolar forces.
   e. $H-O-O-H$ has London dispersion forces, dipolar forces, and hydrogen bonding.
   f. H_2CCl_2 has London dispersion forces and dipolar forces.

2. a. Xe has greater dispersion forces.
   b. H_3C-O-H has H bonding.
   c. H_2CCl_2 is more polar.

3. In the storage bottle, ether is in equilibrium with its vapor and no evaporation takes place; therefore, it is at the same temperature with its surroundings. When ether is poured into an evaporating dish, the higher-energy particles escape, leaving the molecules with lower kinetic energy behind. The average kinetic energy of the remaining particles is less than it was in the bottle and the temperature is therefore lower.

4. $P_{total} = P_O + P_L$ (Dalton's law of partial pressures)

   832 torr = 755 torr + P_L

   P_L = 77 torr

5. $P_T = P_{O_2} + P_{H_2O}$

   Given: P_T = 759 torr P_{H_2O} = 24 torr

   759 torr = P_{O_2} + 24 torr

   P_{O_2} = 735 torr

   $PV = nRT$

   P = 735 torr $R \doteq 62.4 \dfrac{\text{liter torr}}{\text{mole K}}$

$$V = 0.500 \text{ liters} \quad T = 298 \text{ K}$$

$$735 \text{ } \cancel{\text{torr}} \times 0.500 \text{ } \cancel{\text{liters}} = n \times 62.4 \text{ } \frac{\text{liter torr}}{\text{mole K}} \times 298 \text{ } \cancel{K}$$

$$n = 0.0454 \text{ mole}$$

6. $P_T = P_s + P_1$

   P_s = the pressure of the original gas which may be calculated:

$$P_s = P_o \times \frac{V_o}{V_s}$$

$$P_s = 760 \text{ torr} \times \frac{250 \text{ } \cancel{ml}}{276 \text{ } \cancel{ml}}$$

$$P_s = 688 \text{ torr}$$

$$P_T = P_s + P_1$$

$$760 \text{ torr} = 688 \text{ torr} + P_1$$

$$P_1 = 72 \text{ torr}$$

7. a. Lowest SiH_4 , SiF_4 , $SiCl_4$, $SiBr_4$; SiI_4 highest. The increased molecular weight and size results in increased London dispersion forces.
   b. Lowest H_3C-CH_3 , H_3CF ; H_3C-O-H highest.

SECTION 9.8 - 9.12 Types of Solids

8. Classify the following solids as ionic, molecular, covalent, or metallic from their formula and melting point.
   a. SiS_2 , mp = 1090°C
   b. SeO_2 , mp = 340°C
   c. KF, mp = 846°C
   d. Sn, mp = 232°C
   e. Ag , mp = 961°C
   f. AlF_3 , 1291°C
   g. Si_3N_4 , mp = 1900°C
   h. SiF_4 , mp = -90.2°C
   i. SiC , mp = 2700°C

SECTION 9.15

9. Calculate the heat required to convert 1 kg of ethyl

alcohol, C_2H_6O , from liquid to vapor at its boiling point if the molar heat of vaporization is 9.67 kcal/mole.

10. Calculate the heat required to convert 1 kg of H_2O from solid at $-10^\circ C$ to steam at $110^\circ C$.
    molar heat of fusion = 1.435 kcal/mole
    molar heat of vaporization = 9.72 kcal/mole
    heat capacity $H_2O_{(s)}$ = 0.5 cal/g$^\circ$C
    heat capacity $H_2O_{(1)}$ = 1.0 cal/g$^\circ$C
    heat capacity $H_2O_{(g)}$ = 0.5 cal/g$^\circ$C

11. Calculate the heat required to convert 1 m^3 of $H_2O_{(1)}$ to $H_2O_{(g)}$ if the density of H_2O is 1 g/ml.

12. Calculate the heat capacity of Zn given the fact that when 100 g Zn at $90^\circ C$ are added to 100 g of H_2O at $20^\circ C$, the resulting temperature is $26^\circ C$.

Answers:

8. a. SiS_2 is a compound containing only nonmetals and therefore is not ionic. Its high melting point would indicate that it is a covalent solid.
   b. molecular solid
   c. ionic solid
   d. metallic solid
   e. metallic solid
   f. ionic solid
   g. covalent solid
   h. molecular solid
   i. covalent solid

9. Given: 1000 g C_2H_6O
              molar heat of fusion = 9.67 kcal/mole

   1 000 g C_2H_6O x $\dfrac{1 \text{ mole } C_2H_6O}{18 \text{ g } C_2H_6O}$ x 9.67 kcal/mole C_2H_6O

   = 167 kcal

10. Step 1. $H_2O_{(s)}$ - $10^\circ C$ → $H_2O_{(s)}$ $0^\circ C$

              1 000 g x 0.5 cal/g $^\circ C$ x $10^\circ C$ =   5 000 cal

    Step 2. $H_2O_{(s)}$ $0^\circ C$ → $H_2O_{(1)}$ $0^\circ C$

$$1\ 000 \text{ g } H_2O \times \frac{1 \text{ mole } H_2O}{18 \text{ g } H_2O}$$

$$\times 1440 \text{ cal/mole} = \quad 79\ 200 \text{ cal}$$

Step 3. $H_2O_{(1)}$ $0^\circ C \rightarrow H_2O_{(1)}$ $100^\circ C$

$$1\ 000 \text{ g } H_2O \times (1 \text{ cal/g}^\circ C) \times 100^\circ C = 100\ 000 \text{ cal}$$

Step 4. $H_2O_{(1)}$ $100^\circ C \rightarrow H_2O_{(g)}$ $100^\circ C$

$$1\ 000 \text{ g } H_2O \times \frac{1 \text{ mole } H_2O}{18 \text{ g } H_2O}$$

$$\times 9\ 720 \text{ cal/mole} = \quad 540\ 000 \text{ cal}$$

Step 5. $H_2O_{(g)}$ $100^\circ C \rightarrow H_2O_{(g)}$ $110^\circ C$

$$1\ 000 \text{ g } H_2O \times (0.5 \text{ cal/g}^\circ C) \times 10^\circ C = \underline{5\ 000 \text{ cal}}$$

$$\text{Total} = 729\ 200 \text{ cal}$$

11. $1 \text{ m}^3 \times \dfrac{1 \times 10^6 \text{ cm}^3}{1 \text{ m}^3} \times 1 \dfrac{g}{\text{cm}^3} = 1 \times 10^6 \text{ g } H_2O$

$1 \times 10^6 \text{ g } H_2O \times \dfrac{1 \text{ mole } H_2O}{18 \text{ g } H_2O} \times 9.72 \dfrac{\text{kcal}}{\text{mole}} = 5.40 \times 10^5 \text{ kcal}$

12. Heat gain = Heat loss

$100 \text{ g } H_2O \times 1 \dfrac{\text{cal}}{g^\circ C} \times 6^\circ C = 100 \text{ g} \times \text{Heat capacity} \times 64^\circ C$

$600 \text{ cal} = 6400 \text{ g}^\circ C \times \text{Heat capacity}$

$\text{Heat capacity} = \dfrac{600 \text{ cal}}{6400 \text{ g}^\circ C} = 0.094 \dfrac{\text{cal}}{g^\circ C}$

SELF-TEST 1. The attractive forces that exist between all molecules are called _____.
2. In order for a substance to have hydrogen bonding
between molecules, the substance must contain what
atomic groupings?
3. Which of the attractive forces arises from the difference in electronegativity between atoms and the
geometry of molecules?
4. Which of the attractive forces between molecules is

dependent on the molecular size and complexity of the molecules?

5. AlF_3 melts at 1281^oC while $AlBr_3$ melts at 97^oC. What does this suggest about the two substances?

6. List the following substances in order of increasing boiling points: H_3CF , H_3CBr , H_3CI , H_3CCl.

7. List the attractive forces that exist between molecules of $H_3C-\underset{H}{N}-H$.

8. Explain why a volatile liquid becomes cooler when it is poured into an open dish.

9. If a given volume of a pure gas is bubbled through water, what will happen to the volume of the gas?

10. A small quantity of a liquid is injected into a system with an expansible volume that has a pressure of 750 torr and a temperature of 20^oC. What is the vapor pressure of the liquid if the volume of the container expands from 1.0 liter to 1.10 liter?

11. Why will food cook faster if it is boiled in a pressure cooker?

12. In general, which type of solid has the lowest melting point?

13. When a substance boils or evaporates, heat is put into the system with no increase in temperature. Where does the energy go?

14. If the heat of fusion for water is 1.44 kcal/mole, what quantity of heat must be removed to freeze 500 g of H_2O?

15. Thunderstorms usually contain a large amount of energy that has been released from the system. Where does this energy come from?

16. What is the binding force in a crystal of KBr?

17. The partial pressure of water in a 1 liter container with 250 ml of liquid water at 25^oC is 24 torr. What is the partial pressure of water if the volume of the container is increased to 2 liters?

The following problem pertains to Questions 18 - 20: 1800 g of water are heated from 25^oC to form steam at 120^oC. The heat capacity of $H_2O_{(1)}$ is 1 cal/g ^oC; the heat capacity of $H_2O_{(g)}$ is 0.5 cal/g ^oC; and the heat of vaporization of water is 9 720 cal/mole.

18. What quantity of heat is required to heat the liquid water to its boiling point?

19. What quantity of heat is required to convert the

water to steam?
20. What quantity of heat is required to heat the steam
to 120°C?

Answers:

1. London dispersion forces
2. N-H , O-H , or H-F
3. Dipole-dipole attractions
4. London dispersion forces
5. AlF_3 is an ionic solid and $AlBr_3$ is a molecular
solid
6. H_3CF , H_3CCl , H_3CBr , H_3CI
7. All three types, London dispersion forces, dipolar
attractions, and hydrogen bonding.
8. As the volatile liquid evaporates, it is the higher-
energy molecules that escape leaving the molecules
with lower kinetic energy behind. This results in
the decrease of the temperature of the liquid.
9. The volume increases because it becomes saturated
with water vapor.
10. 68 torr
11. In a pressure cooker the pressure on the liquid is
considerably above the atmospheric pressure. This
increase in pressure increases the boiling point of
the liquid and the food cooks faster because of the
increased temperature.
12. Molecular solids
13. The energy is used to overcome the attractive forces
between the molecules and is transformed into potential energy that is recovered when the process is
reversed.
14. 40 kcal
15. The energy comes from the condensation of the water
vapor into rain. (See the answer for Question 13.)
16. Ionic bonds
17. 24 torr, the equilibrium vapor pressure does not
change
18. 135 000 cal
19. 972 000 cal
20. 36 000 cal

chapter

10 Solutions

OBJECTIVES
1. You should be familiar with the terms for the components of a solution (Section 10.1).
2. You should be able to list the various types of solutions and be able to give examples of each type (Section 10.2).
3. You should be able to list the properties of solutions (Section 10.3).
4. You should know the general rule for solubility (Section 10.4).
5. You should be able to describe the intermolecular attractive forces that affect the solubility of various solutes in various solvents (Section 10.4).
6. You should know the interactions that take place between water as a solvent and ionic solutes (Section 10.4).
7. You should be able to explain what is meant by an unsaturated solution, a saturated solution, and a supersaturated solution (Section 10.5).
8. You should be able to explain the simple weight-volume method of expressing concentrations of solutions (Section 10.5).
9. You should be able to explain how to prepare solutions based on molarity (Section 10.6).
10. You should be able to calculate the concentration in molarity from the weight of the solute and the volume of the solution (Section 10.6).
11. You should be able to calculate the moles of solute and weight of solute from the given molarity of a solution and the volume of the solution (Section

10.6).

12. You should be able to calculate the molarity of a solution after a dilution process from the molarity and volume of the original solution (Section 10.6).

13. You should be able to calculate the weight concentration of a solution (Section 10.7).

14. You should be able to calculate the molarity of a solution from the weight % and the density of a solution (Section 10.7).

15. You should be able to describe how you would prepare a solution of a given molality (Section 10.8).

16. You should be able to calculate the molality from any given weight of solute and any given weight of solvent (Section 10.8).

17. You should be able to calculate the molality of a solution from the weight % concentration (Section 10.8).

18. You should be able to explain what is meant by concentration expressed as parts per million or parts per billion (Section 10.10).

19. You should be able to explain what is meant by percent by volume (Section 10.9).

20. You should be able to calculate the volume of solutions required to give the necessary quantity of reactants for a given reaction (Section 10.11).

21. You should be able to calculate the change in freezing point of a molal solution (Section 10.12).

22. You should be able to determine the molecular weight of an unknown solute from the weights of the solute and solvent and the freezing point of the solution (Section 10.12).

23. You should be able to determine the effect of a solute on the boiling point of the solvent (Section 10.13).

24. You should be able to calculate the increase in the boiling point of a solvent caused by a nonvolatile solute (Section 10.13).

25. You should be able to describe the process of osmosis (Section 10.14).

26. You should be able to calculate the osmotic pressure of a solution from the concentration (Section 10.14).

27. You should be able to describe the significant difference between a solution and a colloidal suspension (Section 10.15).

28. You should be able to list the kinds of colloidal dispersions (Section 10.14).

Section 10.1

Solution   a homogeneous form of matter having a variable composition. Solutions are a form of matter in which one or more substances are dispersed in another substance on a molecular-particle scale of atoms, ions, or molecules.

Solute   the component of a solution which is dispersed in another substance.

Solvent   the substance in which other substances are dispersed to form a solution.

Section 10.2

Solutions are usually classified by the physical states of the components in their natural state. See Table 10.1, page 346 of the text for a summary of the types of solutions.

Section 10.4

General rule of solubility   likes dissolve likes.

Hydration   the interaction of water as a solvent with the particles (ions or molecules) of the solute.

Section 10.5

Unsaturated solution   a solution that contains less than the amount of solute required to be in equilibrium with undissolved solute.

Saturated solution   a solution that contains the amount of solute required to be in equilibrium with undissolved solute.

Supersaturated solution   a solution that contains more than the amount of solute required to be in equilibrium with undissolved solute.

Weight-volume concentration   the weight of solute in a given volume of solvent. In the notation,

$$\text{Solubility of KBr} = 53.48^0 \text{ g/100 ml } H_2O$$

$$= 102^{100} \text{ g/100 ml } H_2O$$

the superscripts 0 and 100 are the temperatures at

which the given solubilities of a saturated solution were measured.

Section 10.6

Molarity (M)   the concentration measured in moles of solute in 1 liter of solution.

$$M = \frac{\text{moles solute}}{\text{liter solution}}$$

Section 10.7

Percent by weight concentration (wt.%)   the percent solute of the total solution.

$$\text{wt.\%} = \frac{\text{wt. solute}}{\text{wt. solution}} \times 100$$

Section 10.8

Molality (m)   the concentration of a solution expressed as moles of solute in 1 000 g of solvent.

$$m = \frac{\text{moles solute}}{1\,000 \text{ g solvent}}$$

Section 10.9

Percent by volume   a method of expressing concentration of liquid - liquid solutions in which the % of the volume made up of solute is used.

Section 10.10

Parts per million (ppm)   the concentration expressed as a ratio of the weight of solute to 1 000 000 parts by weight solution.

Parts per billion (ppb)   the concentration expressed as a ratio of the weight of solute to 1 000 000 000 parts by weight solution.

Section 10.11

See the problems for this Section.

Section 10.12

Molal <u>freezing</u> <u>point</u> <u>constant</u>   the change in the
freezing point of a solvent caused by the presence
of 1 mole of solute particles in 1 000 g of solvent.
The constant is generally independent of the type of
solute particle. In the formula

$$\Delta T_f = K_f \times m$$

ΔT_f is the change in the freezing point of the solvent; K_f is the freezing point constant.

See the problems for this section including molecular
weight determinations by freezing point depression.
The freezing point of all solvents is depressed by
the presence of a solute or impurity.

Section 10.13

Molal <u>boiling</u> <u>point</u> <u>constant</u> <u>(K_b)</u>   the change in the
boiling point of a solvent caused by the presence of
1 mole of nonvolatile solute particles in 1 000 g of
solvent. The solute must be nonvolatile if the expression: $\Delta T_b = m \times K_b$ is valid. The effect of
volatile solute on the boiling point of a solution
is more complex and will not be considered at this
time.

Section 10.14

Osmosis   the process by which a solvent such as water
flows from a less-dense to a more-dense solution
through a membrane that is permeable only to the
solvent particles.

Osmotic <u>pressure</u>   the pressure that must be applied to
the less-dense solution on an osmotic process in
order to have the rate of flow of solvent particles
the same in both directions through a semipermeable
membrane. In the formula

$$\pi = MRT$$

π = osmotic pressure; R = ideal gas constant; M =
molarity of solution; T = absolute temperature.

SECTION 10.15

Colloid   a state of matter in which finely divided
    particles of one substance (the dispersed phase) are
    suspended in another substance (the dispersing med-
    ium).

Aerosol   a colloid consisting of a solid or a liquid
    dispersed in a gas.

Sol   a colloid consisting of a solid dispersed in a
    liquid.

Emulsion   a colloid consisting of a liquid dispersed
    in another liquid.

Gel   a colloid consisting of a liquid dispersed in a
    solid. Colloidal particles or the dispersed phase
    particles range in size from diameters of 1×10^{-4}
    to 1×10^{-7} cm.

Tyndall effect   the reflection or scattering of light
    by the dispersed phase particles of a colloid.

Brownian motion   the random motion of colloidal parti-
    cles caused by collisions with molecules of the dis-
    persing medium.

QUESTIONS
AND
PROBLEMS

SECTION 10.1 - 10.2   What Are Solutions and Types of
                       Solutions

1. Consider the following solutions and classify them,
   identifying the solvent and solute particles.
   a. potassium bromide in water
   b. sugar ($C_{12}H_{22}O_{11}$) and water
   c. gasoline
   d. air
   e. sterling silver (see a dictionary or handbook for
      composition)
   f. zinc amalgam
   g. champagne

SECTION 10.4   General Rules of Solubility

2. Comment on the statement: All mixtures of gases are
   solutions.

3. Predict the solubility of the following substances
   in water:
   a. ammonia, NH_3

b. NaCl
c. ethylene glycol, $H-O-CH_2CH_2-O-H$
d. HF
e. natural gas, CH_4
f. $H_2S_{(g)}$
g. $CF_4{(g)}$

4. Why are the following substances water soluble:
$H_3C-\underset{\underset{H}{|}}{N}-H$, H_3C-O-H , HF , $H_3C-\underset{\underset{CH_3}{|}}{N}-CH_3$, $H_3C-O-CH_3$.

Answers:

1. a. Solid-liquid solution, water is the solvent,
   potassium ions and bromide ions are the solute
   particles.
   b. Solid-liquid solution, water is the solvent, the
   solute particles are sugar molecules.
   c. Liquid-liquid solution, hydrocarbon molecules are
   the solute particles.
   d. Gas-gas solution, molecules of nitrogen and oxygen are the solution particles.
   e. Solid-solid solution, copper atoms are the solute
   particles, silver is the solvent.
   f. Solid-liquid solution, zinc atoms are the solute
   particles and mercury is the solvent.
   g. Gas-liquid-liquid solution, CO_2 and ethanol molecules are the solute particles.

2. Gases consist of widely separated molecules that
   have very small attractive forces between molecules
   in comparison to their kinetic energy. Since solutions are molecular dispersions, mixtures of gases
   are solutions.

3. All of these substances except for e. natural gas,
   CH_4 , are water soluble. g. $CF_4{(g)}$ is least soluble
   of the others but the hydrogen of water can hydrogen
   bond to some extent with the fluorines of carbon
   tetrafluoride.

4. The first three substances have mutual hydrogen
   bonding with water. The latter two will hydrogen
   bond with the hydrogen of water even though they
   have no hydrogen bonding hydrogen themselves.

Section 10.6   Molarity

5. Explain how you would prepare:
   a. 1 liter of a 2.50 M solution of NaCl
   b. 500 ml of a 0.1000 M AgNO$_3$ solution

6. Calculate the molarities of the following solutions:
   a. 26.13 g of Ba(NO$_3$)$_2$ in 1 liter of solution
   b. 2.000 g CaBr$_2$ in 50 ml of solution
   c. 50.0 g K$_3$PO$_4$ in 500 ml of solution
   d. 10.20 g Na$_2$C$_2$O$_4$ in 250 ml of solution
   e. 6.000 g HF in 200 ml of solution

7. Calculate the volume of each of the following solutions required to provide the stated quantity of solute:
   a. 1 mole of NaOH from a 6.00 M solution of NaOH
   b. 0.250 mole HCl from a 3.00 M solution
   c. 0.250 mole of KCl from a 5.0 M solution
   d. 5 mole of H$_2$SO$_4$ from a 12 M solution
   e. 10 moles of HCl from a 6.00 M solution

8. What weight of solute is contained in the stated volume of each of the following solutions:
   a. 0.500 liter of a 6.00 M H$_2$SO$_4$ solution
   b. 25 ml of a 0.1250 M AgNO$_3$ solution
   c. 100 ml of a 0.500 M NaOH solution
   d. 2.500 liters of a 12.0 M H$_2$SO$_4$ solution
   e. 1.00 ml of a 0.100 M NaCl solution

9. What volume of a 12.0 M solution of H$_2$SO$_4$ must be used to prepare 0.500 liter of 3.00 M H$_2$SO$_4$?

10. What volume of a 10.0 M solution of NaOH must be used to prepare 1.00 liter of 2.50 M NaOH?

Answers:

5. a. $\dfrac{2.50 \; \text{mole}}{1 \; \text{liter}}$ x $\dfrac{58.5 \; \text{g NaCl}}{1 \; \text{mole NaCl}}$ x 1.00 liter = 146 g NaCl

   Place 146 g of NaCl in a 1 liter volumetric flask and add enough water to make a liter of solution.

   b. $\dfrac{0.100 \; \text{mole}}{1 \; \text{liter}}$ x $\dfrac{170 \; \text{g AgNO}_3}{1 \; \text{mole AgNO}_3}$ x 0.500 liter = 8.50 g

Place 8.50 $AgNO_3$ in a 500 ml volumetric flask and add enough water to make 500 ml of solution.

6. a. $26.13 \text{ g BaCl}_2 \times \dfrac{1 \text{ mole BaCl}_2}{208 \text{ g BaCl}_2} = 0.126 \text{ mole BaCl}_2$

$\dfrac{0.126 \text{ mole}}{1 \text{ liter}} = 0.126 \text{ M}$

b. $2.000 \text{ g CaBr}_2 \times \dfrac{1 \text{ mole CaBr}_2}{200 \text{ g CaBr}_2} = 0.0100 \text{ mole CaBr}_2$

$\dfrac{0.0100 \text{ mole}}{50 \text{ ml}} \times \dfrac{1000 \text{ ml}}{1 \text{ liter}} = 0.200 \text{ M}$

c. 0.471 M

d. 0.304 M

e. 1.50 M

7. a. $\dfrac{1 \text{ mole}}{6.00 \text{ mole/liter}} = 1 \text{ mole} \times \dfrac{1 \text{ liter}}{6.00 \text{ mole}} = 0.167 \text{ liter}$

b. $\dfrac{0.250 \text{ mole}}{3.00 \text{ mole/liter}} = 0.250 \text{ mole} \times \dfrac{1 \text{ liter}}{3.00 \text{ mole}}$

$= 0.0833 \text{ liter}$

c. 0.050 liter

d. 0.417 liter

e. 1.67 liter

8. a. $\dfrac{6.00 \text{ mole}}{1 \text{ liter}} \times 0.500 \text{ liter} \times \dfrac{98.0 \text{ g}}{1 \text{ mole}} = 294 \text{ g}$

b. $\dfrac{0.1250 \text{ mole}}{1 \text{ liter}} \times 25 \text{ ml} \times \dfrac{1 \text{ liter}}{1000 \text{ ml}} \times \dfrac{170.0 \text{ g}}{1 \text{ mole}} = 0.5312 \text{ g}$

c. 2.00 g

d. 2940 g

e. 0.00584 g

9. Given: M_0 = 12 M M_f = 3.0 M
            v_0 = ? v_f = 0.500 liter

$$M_0 \times v_0 = M_f \times v_f$$

$$12\ \frac{mole}{liter} \times v_0 = 3.0\ \frac{mole}{liter} \times 0.500\ liter$$

$$v_0 = 3.0\ \frac{\cancel{mole}}{\cancel{liter}} \times \frac{1\ \cancel{liter}}{12\ \cancel{mole}} \times 0.500\ liter$$

$$v_0 = 0.125\ liter$$

10. 0.250 liter

SECTION 10.7   Weight-Percent Solution

11. What weight of solute is contained in each of the following solutions:
    a. 1040 g of 7.00% $(NH_4)_2SO_4$
    b. 1 liter of a 10.0% $BaCl_2$ solution, density 1.092 g/ml
    c. 500 ml of 50.00% C_2H_6O solution, density 0.9150 g/ml
    d. 3 kg of a 28.00% HCl solution
    e. 100 ml of a 10.00% $MgSO_4$ solution, density 1.105 g/ml

12. Calculate the molarity of the following solutions:
    a. 20.00% HNO_3 , density 1.1170 g/ml
    b. 72.00% H_2SO_4 , density 1.6367 g/ml
    c. 20.00% NaOH , density 1.2214 g/ml
    d. 25.00% NaCl , density 1.191 g/ml
    e. 38.00% HNO_3 , density 1.2357 g/ml

Answers:

11. a. $1040\ g \times \frac{7.00\ g}{100\ g} = 72.8\ g$

    b. $1.092\ \cancel{g/ml} \times \frac{1000\ \cancel{ml}}{1\ \cancel{liter}} \times 1\ \cancel{liter} \times \frac{10.0\ g}{100\ \cancel{g}} = 109.2\ g$

    c. 229 g

    d. 840 g

e. 11.05 g

12. a. $\dfrac{1.1170 \text{ g}}{1 \text{ ml}}$ x $\dfrac{1000 \text{ ml}}{1 \text{ liter}}$ x $\dfrac{20.00 \text{ g}}{100 \text{ g}}$ x $\dfrac{1 \text{ mole } HNO_3}{63.0 \text{ g } HNO_3}$

        ⏜   ⏜   ⏜

    calculate the   convert   convert wt.
    weight of 1 liter   from %   to moles

      = $\dfrac{3.55 \text{ mole}}{1 \text{ liter}}$ = 3.55 M

b. $\dfrac{1.637 \text{ g}}{1 \text{ ml}}$ x $\dfrac{1000 \text{ ml}}{1 \text{ liter}}$ x $\dfrac{72.0 \text{ g}}{100 \text{ g}}$ x $\dfrac{1 \text{ mole } H_2SO_4}{98.0 \text{ g } H_2SO_4}$

      = $\dfrac{12.0 \text{ mole}}{1 \text{ liter}}$ = 12.0 M

c. 6.10 M

d. 5.10 M

e. 7.45 M

SECTION 10.8   Molality

13. Calculate the molality of the following solutions
    that contain:
    a. 46 g NaCl in 500 g H_2O
    b. 360 g $C_6H_{12}O_6$ in 250 g H_2O
    c. 100 g $NaSO_4$ in 1000 g H_2O
    d. 250 g $BaCl_2$ in 500 g H_2O

14. Calculate the molality of the following solutions:
    a. 20.00% HNO_3
    b. 72.00% H_2SO_4
    c. 20.00% NaOH
    d. 25.00% NaCl
    e. 38.00% HNO_3

Answers:

13. a. $\dfrac{46.0 \text{ g NaCl}}{500 \text{ g } H_2O}$ x $\dfrac{1000 \text{ g } H_2O}{1000 \text{ g } H_2O}$ x $\dfrac{1 \text{ mole NaCl}}{58.4 \text{ g NaCl}}$

                   ⏜   ⏜

    determine the quantity   convert wt.
    of solute in 1000 g   to moles
    solvent

$$= \frac{1.58 \text{ mole NaCl}}{1000 \text{ g } H_2O} = 158 \text{ m}$$

b. $\dfrac{360 \text{ g } C_6H_{12}O_6}{250 \text{ g } H_2O} \times \dfrac{1000 \text{ g } H_2O}{1000 \text{ g } H_2O} \times \dfrac{1 \text{ mole } C_6H_{12}O_6}{180 \text{ g } C_6H_{12}O_6}$

$$= \frac{8.00 \text{ mole } C_6H_{12}O_6}{1000 \text{ g } H_2O} = 8.00 \text{ m}$$

c. 0.704 m

d. 2.40 m

14. a. In a 20.0% solution of HNO_3 , there are 20.0 g HNO_3 and 80.0 g H_2O .

$$\frac{20.0 \text{ g } HNO_3}{80.0 \text{ g } H_2O} \times \frac{1000 \text{ g } H_2O}{1000 \text{ g } H_2O} \times \frac{1 \text{ mole } HNO_3}{63.0 \text{ g } HNO_3}$$

$\underbrace{\qquad\qquad\qquad\qquad}$ $\underbrace{\qquad\qquad}$

determine the quantity   convert wt.
of solute in 1000 g   to moles
solvent

$$= \frac{3.97 \text{ mole } HNO_3}{1000 \text{ g } H_2O} = 3.97 \text{ m}$$

b. 26.2 m

c. 6.25 m

d. 5.71 m

e. 9.73 m

SECTION 10.11   Stoichiometry of Solutions

15. What weight of Zn will dissolve in 250 ml of 3.00 M HCl ?

$$Zn + 2 \text{ HCl} \rightarrow ZnCl_2 + H_2$$

16. What volume of 10.0 M acetic acid $H(C_2H_3O_2)$ is required to react with 10.0 g $CaCO_3$?

$$CaCO_3 + 2 \text{ H}(C_2H_3O_2) \rightarrow Ca(C_2H_3O_2)_2 + H_2O + CO_2$$

17. What volume of 3.00 M HCl will be required to react with Zn to produce 5.00 moles of $H_{2(g)}$?

$$Zn + 2 \text{ HCl} \rightarrow ZnCl_2 + H_2$$

Answers:

15. $0.250 \; \cancel{\text{liter}} \; \times \; \dfrac{3 \; \text{mole HCl}}{1 \; \cancel{\text{liter}}} = 0.75 \; \text{mole HCl}$

   $0.75 \; \cancel{\text{mole HCl}} \; \times \; \dfrac{1 \; \cancel{\text{mole Zn}}}{2 \; \cancel{\text{mole HCl}}} \; \times \; \dfrac{65.4 \; \text{g Zn}}{1 \; \cancel{\text{mole Zn}}} = 24.5 \; \text{g Zn}$

16. $10.0 \; \cancel{\text{g CaCO}_3} \; \times \; \dfrac{1 \; \cancel{\text{mole CaCO}_3}}{100 \; \cancel{\text{g CaCO}_3}} \; \times \; \dfrac{2 \; \text{mole H(C}_2\text{H}_3\text{O}_2)}{1 \; \cancel{\text{mole CaCO}_3}}$

   $= 0.20 \; \text{mole H(C}_2\text{H}_3\text{O}_2)$

   $\dfrac{0.20 \; \text{mole H(C}_2\text{H}_3\text{O}_2)}{10.0 \; \text{mole/liter}} = 0.20 \; \cancel{\text{mole}} \; \times \; \dfrac{1 \; \text{liter}}{10.0 \; \cancel{\text{mole}}} = 0.020 \; \text{liter}$

17. $5 \; \cancel{\text{mole H}_2} \; \times \; \dfrac{2 \; \text{mole HCl}}{1 \; \cancel{\text{mole H}_2}} = 10.0 \; \text{mole HCl}$

   $\dfrac{10.0 \; \text{mole HCl}}{3.00 \; \text{mole/liter}} = 10.0 \; \cancel{\text{mole HCl}} \; \times \; \dfrac{1 \; \text{liter}}{3.00 \; \cancel{\text{mole}}} = 3.33 \; \text{liter}$

SECTION 10.12 Lowering of the Solvent Freezing Point

18. What is the freezing point of a 5.0 molal solution of ethanol in H_2O? ($K_f = -1.86^\circ C$.)

19. What weight of ethylene glycol (antifreeze), $C_2H_6O_2$, must be added to 2.00 kg of H_2O in order to have a solution with a freezing point below $-20^\circ C$? ($K_f = -1.86^\circ C$.)

20. A 1.00 g sample of an unknown substance was dissolved in 10 g of cyclohexane. The resulting solution had a freezing point of $-20^\circ C$. What is the molecular weight of the substance? (Cyclohexane: fp $= 6.5^\circ C$, $K_f = -20^\circ C$.)

21. A 0.750 g sample of an unknown substance was dissolved in 10.0 g of cyclohexane. The resulting solution had a freezing point of $-7.2^\circ C$. What is the molecular weight of the unknown substance? (Cyclohexane: fp $= 6.5^\circ C$, $K_f = -20^\circ C$.)

SECTION 10.14 Osmosis

22. Calculate the height to which a column of a solution can be raised by the osmotic pressure between water

and a 0.025 M sugar solution at 25°C , if a 10.4 m
column of the solution exerts a pressure of 1 atmosphere.

Answers:

18. $\Delta T_f = K_f \text{ x m}$ Given: m = 5; $K_f = -1.86^{\circ}$C.

    $\Delta T_f = (-1.86^{\circ}\text{C}) \text{ x } 5$

    $\Delta T_f = -9.3^{\circ}$C

    $0^{\circ}\text{C} - 9.3^{\circ}\text{C} = -9.3^{\circ}\text{C}$ = freezing point

19. $\Delta T_f = K_f \text{ x m}$ Given: $\Delta T_f = -20^{\circ}$C ; $K_f = -1.86^{\circ}$C.

    $-20^{\circ}\text{C} = -1.86^{\circ}\text{C x m}$

    m = 10.8

    $$\frac{10.8 \text{ mole } C_2H_6O_2}{1000 \text{ g } H_2O} \text{ x } 2\,000 \text{ g } H_2O \text{ x } \frac{62 \text{ g } C_2H_6O_2}{1 \text{ mole } C_2H_6O_2}$$

    $= 1\,290 \text{ g } C_2H_6O_2$

20. Given: $\Delta T_f = -26.5^{\circ}$C ; $K_f = -20^{\circ}$C

    $\Delta T_f = K_f \text{ x m}$

    $-26.5^{\circ}\text{C} = -20^{\circ}\text{C x m}$

    m = 1.32

    $$\frac{1.32 \text{ mole}}{1000 \text{ g}} \text{ x } 10 \text{ g} = 0.0132 \text{ mole}$$

    1.00 g = 0.0132 mole

    75.8 g = 1 mole

21. 109 g

22. Given: M = 0.025 ; T = 298 K ; R = 0.082 $\dfrac{\text{liter atm}}{\text{mole K}}$

    $\pi = MRT$

$$\pi = 0.025 \; \frac{\text{mole}}{\text{liter}} \; \times \; 0.082 \; \frac{\text{liter atm}}{\text{mole K}} \; \times \; 298 \; \text{K}$$

$$\pi = 0.61 \text{ atm}$$

$$0.61 \text{ atm} \times \frac{10.4 \text{ m}}{1 \text{ atm}} = 6.4 \text{ m}$$

SELF-TEST
1. The substance that is dissolved is called the
   _____.
2. Concentration expressed as moles of solute per liter
   of solution is _____.
3. The process in which solvent molecules flow through
   a semipermeable membrane from the less-concentrated
   solution to the more-concentrated solution is called
   _____.
4. The dispersion of a finely divided phase in another
   phase is called a _____.
5. Concentration expressed as moles of solute in 1000 g
   of solvent is called _____.
6. The solubility of gases usually _____ with
   an increase in temperature.
7. The solubility of solids usually _____
   with an increase in temperature.
8. The freezing point of a solution is usually
   _____ than the freezing point of the pure
   solvent.
9. What volume of a 6.0 M solution is required to prepare 500 ml of a 0.100 M HCl solution?
10. What weight of Na_3PO_4 is required to prepare 250 ml
    of a 0.500 M solution?
11. What is the molarity of a 4.00 wt.% NaCl solution
    with a density of 1.020 g/ml?
12. What is the molality of a 4.00 wt.% solution of
    NaCl?
13. What is the freezing point of a 50 wt.% solution of
    $C_2H_6O_2$ in water? ($K_f = -1.86°C$.)
14. When a 1.50 g sample of an unknown substance is dis-
    solved in 20.0 g of cyclohexane, the freezing point
    of the solution is $-3.5°C$. The normal freezing
    point of cyclohexane is $6.5°C$ and K_f is $-20°C$. What
    is the molecular weight of the unknown substance?
15. What volume of a 0.100 M $AgNO_3$ solution is required
    to react with 0.150 g of $BaCl_2$ in the reaction:

$$2 \text{ AgNO}_3 + \text{BaCl}_2 \rightarrow 2 \text{ AgCl} + \text{Ba(NO}_3)_2$$

16. What volume of a 6.00 M HNO_3 solution is required to dissolve 0.635 g of Cu in the reaction:

$$8 \text{ HNO}_3 + 3 \text{ Cu} \rightarrow 3 \text{ Cu(NO}_3)_2 + 4 \text{ H}_2\text{O} + 2 \text{ NO}$$

17. Which has the lower freezing point, a 1 molal solution of ethanol, $\text{C}_2\text{H}_6\text{O}$, or a 1 molal solution of CaCl_2 ?
18. What weight of a 5.0 wt.% solution is needed for 12.0 g of $\text{K}_2\text{Cr}_2\text{O}_7$?
19. Brownian motion and the Tyndall effect are phenomena observed in _____ .
20. The headlight beam of cars that is visible in the fog is an example of _____ .

Answers:

1. Solute
2. Molarity (M)
3. Osmosis
4. Colloid
5. Molality
6. Decrease
7. Increases
8. Less
9. 8.3 ml
10. 20.5 g
11. 0.699 M
12. 0.713 m
13. -30°C
14. 150 g
15. 14.4 ml
16. 4.44 ml
17. CaCl_2 , a 1 molal solution is 1 m Ca^{2+}, and 2 m Cl^- or a total of 3 m.
18. 240 g solution
19. Colloids
20. Tyndall effect

11 Aqueous Solutions of Acids, Bases, and Salts

OBJECTIVES

1. You should know the definitions of an acid or a base according to the Arrhenius concept (Section 11.1).
2. You should be able to list the properties of an acid or a base (Section 11.1).
3. You should know the definitions of a Bronsted acid and a Bronsted base (Section 11.2).
4. You should be able to identify acid - base pairs according to the Bronsted concept (Section 11.2).
5. You should be able to write equations for reactions between Bronsted acids and Bronsted bases to form their respective conjugate bases and conjugate acids (Section 11.2).
6. You should be able to write the equation for the ionization of water (Section 11.3).
7. You should be able to write the mathematical expression for the ion product of water (Section 11.4).
8. You should know the ionization constant of water at 25°C as 1×10^{-14} (Section 11.4).
9. Given the hydronium ion concentration, you should be able to calculate the hydroxide ion concentration and vice versa (Section 11.4).
10. You should know the exact meaning of pH and pOH (Section 11.5).
11. Given any one of the values for pH, pOH, $[H_3O^+]$, or $[OH^-]$, you should be able to determine the value for the other three (Section 11.5).
12. You should know what is meant by the term "strong acid" and you should be able to list the common

strong acids (Section 11.6).

13. You should know what is meant by the term "weak acid" (Section 11.7).

14. You should be able to write an equation for the dissociation of any weak acid in water (Section 11.7).

15. You should be able to write the dissociation constant expression for any weak acid (Section 11.7).

16. Given any two of the following values, you should be able to calculate the third: K_a ; concentration of the acid [HA] ; or $[H_3O^+]$ (Section 11.7).

17. You should be able to write the equation for the dissociation or hydrolysis of any weak base in water (Section 11.8).

18. You should be able to write the ionization constant expression for any weak base (Section 11.8).

19. Given the value of any two of the quantities, K_b , base. or $[OH^-]$, you should be able to calculate the value of the third (Section 11.8).

20. You should know what is meant by the term buffer and be able to calculate the pH of a buffer (Section 11.9).

21. You should know which oxides are acidic oxides and you should be able to write the equation for their reactions with water (Section 11.10).

22. You should know what is meant by a basic oxide and be able to write an equation for the reaction of a basic oxide with water (Section 11.10).

23. You should be able to solve problems involving the stoichiometry of acid‐base neutralization reactions (Section 11.11).

24. You should be able to apply the general rules of solubilities of salts in water in order to write equations for double displacement reactions between acids, bases, and salts (Sections 11.12 and 11.13).

25. You should be able to write the ion product expression for any salt and be able to relate it to the solubility product (Section 11.14).

26. Given the values of any two of the terms, solubility product, anion concentration, or cation concentration, you should be able to calculate the third (Section 11.14).

IMPORTANT TERMS AND CONCEPTS

SECTION 11.1

Acid (Arrhenius concept)   any substance that releases hydrogen ions in aqueous solutions.

Base (Arrhenius concept)   any substance that releases hydroxide ions in aqueous solutions.

Common properties of acids:
   a. taste sour
   b. turn blue litmus red
   c. react with carbonates to produce CO_2
   d. react with active metals to produce H_2
   e. neutralize bases

Common properties of bases:
   a. feel slick or soapy
   b. turn red litmus blue
   c. taste bitter
   d. neutralize acids

Section 11.2

Acid (Bronsted concept)   any substance capable of donating a hydrogen ion (H^+) or proton to another substance.

Base (Bronsted concept)   any substance capable of accepting a hydrogen ion (H^+) or proton from another substance.

Bronsted acids and bases are always in conjugate acid-base pairs. The acid has a hydrogen atom bonded to it; the base has lost the hydrogen atom as H^+ or a proton. A Bronsted base always has at least one pair of nonbonded electrons available for accepting a proton.

$$H\text{-}A \xrightarrow{\ -H^+\ } \ |A^-$$
$$\text{acid} \qquad\qquad \text{base}$$

$$H\text{-}\underline{\overline{O}}\text{-}H \longrightarrow \ |\underline{\overline{O}}\text{-}H$$
$$\text{acid} \qquad\qquad \text{base}$$

In acid - base reactions the acid always donates a proton to the base.

$$H\text{-}Cl \ + \ H\text{-}O\text{-}H \ \rightleftharpoons \ Cl^- \ + \ H\text{-}\overset{+}{\underset{\overset{|}{H}}{O}}\text{-}H$$
$$\text{acid I} \quad \text{base II} \qquad \text{base I} \quad \text{acid II}$$

HCl and Cl^- are an acid - base pair. H_3O^+ and H_2O

are an acid-base pair.

Section 11.3

Amphoteric   the ability to undergo a reaction as either an acid or a base. Water is an amphoteric substance; it can react as either a Bronsted acid or a Bronsted base.

H_2O as an acid:

$$H_2O + NH_3 \rightleftharpoons OH^- + NH_4^+$$

H_2O as a base:

$$HCl + H_2O \rightleftharpoons Cl^- + H_3O^+$$

Ionization of water (water is both the Bronsted acid and the Bronsted base):

$$H_2O + H_2O \rightleftharpoons H_3O^+ + OH^-$$

Hydronium ion   the conjugate acid of water as a Bronsted base, i.e., a water molecule that has gained a proton (H^+).

Ion product of water   the product of the concentration of the hydronium ion (or hydrogen ion) times the concentration of the hydroxide ion.

Ion product of water = $[H_3O^+][OH^-]$ or $[H^+][OH^-]$

Ionization constant of water K_w the value of the ion product of water at equilibrium (at 25°C K_w = 1×10^{-14}). At equilibrium: $K_w = [H_3O^+][OH^-] = 1 \times 10^{-14}$.

Neutral solution   an aqueous solution in which the hydronium ion concentration and the hydroxide ion concentration are equal at equilibrium. In a neutral solution, when the hydronium ion concentration is equal to the hydroxide ion concentration, the concentration of each of the ions is 1×10^{-7}M.

$$[H_3O^+] OH^- = 1 \times 10^{-14}$$

$$[H_3O^+] = [OH^-] = 1 \times 10^{-7}$$

Note: Remember the brackets [ ] mean the concentration of the ion in mole/liter or M.

Section 11.5

pH   the negative logarithm (base 10) of the hydronium (or hydrogen) ion concentration in an aqueous solution.

$$pH = -\log\left[H_3O^+\right]$$

Neutral pH   pH = 7.

Acid pH   pH < 7.

Basic pH   pH > 7. The higher the $[H_3O^+]$, the lower the pH. See Table 11.2 in the text for complete pH, pOH, $[H_3O^+]$, $[OH^-]$ relationships.

pOH   the negative log of the hydroxide ion concentration.

$$pOH = -\log [OH^-]$$

Note: pH + pOH = 14. This statement is always true for aqueous solutions at 25°C.

Section 11.6

Strong acid   an acid that is completely ionized or dissociated in aqueous solutions. See Table 11.3 in the text for the common strong acids.

Section 11.7

Weak acid   an acid which is only partly ionized or dissociated in aqueous solutions.

Strong electrolyte   any solute that is a good conductor of electricity (e.g., soluble ionic compounds and strong acids).

Weak electrolyte   any solute that is a poor conductor of electricity (e.g., weak acids and weak nonionic bases).

Nonelectrolyte   any solute that is a nonconductor of electricity (e.g., covalent compounds which are soluble in water).

Dissociation expression for a weak acid   for any acid HA:

$$\frac{[H_3O^+][A^-]}{[HA]}$$

Dissociation constant for an acid (K_a)   the numerical
value of the dissociation expression at equilibrium.
At equilibrium:

$$K_a = \frac{[H_3O^+][A^-]}{[HA]}$$

SECTION 11.8

Dissociation constant of a weak base (K_b)   at equilibrium:

$$K_b = \frac{[BH^+][OH^-]}{[B]}$$

SECTION 11.9

Buffer   a solution that resists changes of pH. Buf-
fers are usually solutions containing nearly equivalent amounts of weak conjugate acid - base pairs or
a weak acid and the salt of the weak acid.

SECTION 11.10

Acidic oxides   oxides of nonmetals that react with
water to form acids.

Basic oxides   oxides of group IA and group IIA metals
that react with water to form hydroxides.

SECTION 11.11

Neutralization   the reaction of an acid with a base to
produce water and a salt.

Titration   the accurate measurement of the volume of a
known concentration of one of the reagents, either
acid or base, that will just completely react with
an accurately measured volume or weight of another
reagent, acid or base.

End point   the equivalence point in a titration.

SECTION 11.12

The rules of solubility for salts in aqueous solutions

are given in this section of the text. You should
be familiar with these rules.

Section 11.13

See the Questions and Problems for this section.

Section 11.14

Ion <u>product</u> <u>of</u> <u>a</u> <u>salt</u>   the product of the molar concentration of the ions of a salt raised to the math-
ematical product of their respective stoichiometric
coefficients in the balance equation for the ionization of the salt.

<u>Solubility</u> <u>product</u>   the value of the ion product when
the ions of a substance in solution are in equili-
brium with the solid substance.

$$K_{sp} = M^{+n}\ A^{-m}$$

QUESTIONS AND PROBLEMS

Section 11.1   Arrhenius Concept of Acids and Bases

1. a. List five acids and show their dissociation ac-
      cording to the Arrhenius concept.
   b. List five bases and show how each would form
      hydroxide ions in solution.

Section 11.2   Bronsted Concept of Acids and Bases

2. Write the equation for the reaction of the following
   Bronsted acids with water:
   a. $H(C_2H_3O_2)$
   b. H_2S
   c. HSO_3^-
   d. HNO_2

3. Write the equation for the reaction of the following
   Bronsted bases with water:
   a. H_2NNH_2
   b. H_2NOH
   c. HSe^-
   d. CO_3^{2-}

Answers:

1. a. $HCl \rightarrow H^+ + Cl^-$

   $H(C_2H_3O_2) \rightarrow H^+ + C_2H_3O_2^-$

   $H_2Se \rightarrow H^+ + HSe^-$

   $HNO_2 \rightarrow H^+ + NO_2^-$

   $H_2SO_4 \rightarrow 2 H^+ + SO_4^{2-}$

   b. $NaOH \rightarrow Na^+ + OH^-$

   $NH_4OH \rightarrow NH_4^+ + OH^-$

   $LiOH \rightarrow Li^+ + OH^-$

   $KOH \rightarrow K^+ + OH^-$

   $Ba(OH)_2 \rightarrow Ba^{2+} + 2 OH^-$

2. a. $H(C_2H_3O_2) + H_2O \rightleftharpoons H_3O^+ + C_2H_3O_2^-$

   b. $H_2S + H_2O \rightleftharpoons H_3O^+ + HS^-$

   c. $HSO_3^- + H_2O \rightleftharpoons SO_3^- + H_3O^+$

   d. $HNO_2 + H_2O \rightleftharpoons NO_2^- + H_3O^+$

3. a. $H_2NNH_2 + H_2O \rightleftharpoons H_2NNH_3^+ + OH^-$

   b. $H_2NOH + H_2O \rightleftharpoons H_3NOH^+ + OH^-$

   c. $HSe^- + H_2O \rightleftharpoons H_2Se + OH^-$

   d. $CO_3^{2-} + H_2O \rightleftharpoons HCO_3^- + OH^-$

SECTIONS 11.3 - 11.4   Ionization and Ion Product of Water

4. Find the hydroxide ion concentration in each of the following solutions:
   a. 0.001 M HCl
   b. a solution containing 0.0036 g HCl in 1 liter of solution.

c. a solution with a hydronium ion concentration of 1×10^{-9}.

SECTION 11.5 The pH Scale

5. Find the pH of the following solutions:
   a. 0.005 M HCl
   b. 0.005 M NaOH
   c. a solution that has a hydronium ion concentration of 4.5×10^{-5}M.
   d. a solution with a hydroxide ion concentration of 3.16×10^{-6}M.

6. Give the hydronium ion concentration and the hydroxide ion concentration of solutions with:
   a. a pH of 7
   b. a pH of 9.5
   c. a pH of 4.7
   d. a pOH of 3.6
   e. a pOH of 9.2

Answers:

4. a. Given: $[H_3O^+]$ = 0.001 M

   $$[H_3O^+][OH^-] = 1 \times 10^{-14}$$

   $$0.001 \text{ M } [OH^-] = 1 \times 10^{-14}$$

   $$[OH^-] = 1 \times 10^{-11} \text{ M}$$

   b. Given: 0.0036 g HCl

   $$\frac{0.0036 \text{ g HCl}}{1 \text{ liter}} \times \frac{1 \text{ mole HCl}}{36 \text{ g HCl}} = 1 \times 10^{-4} \text{ M HCl}$$

   $$[H_3O^+] = 1 \times 10^{-4} \text{ M}$$

   $$[H_3O^+][OH^-] = 1 \times 10^{-14}$$

   $$[OH^-] = 1 \times 10^{-10}$$

   c. Given: $[H_3O^+] = 1 \times 10^{-9}$ M

   $$[H_3O^+][OH^-] = 1 \times 10^{-14}$$

   $$[OH^-] = 1 \times 10^{-5} \text{ M}$$

5. a. Given: 0.005 M HCl

$$[H_3O^+] = 0.005 \text{ M}$$

$$pH = -\log [H_3O^+]$$

$$pH = -\log 0.005$$

$$pH = 2.3$$

Note: You should review the section on logarithms in Chapter 2. The use of a scientific pocket calculator is recommended in working with logarithms.

b. Given: $[OH^-] = 0.005$ M

$$[H_3O^+][OH^-] = 1 \times 10^{-14}$$

$$[H_3O^+] = 2 \times 10^{-12}$$

$$pH = -\log[H_3O^+]$$

$$pH = -\log 2 \times 10^{-12}$$

$$pH = 11.7$$

c. Given: $[H_3O^+] = 4.5 \times 10^{-5}$ M

$$pH = -\log[H_3O^+]$$

$$pH = 4.3$$

d. Given: $[OH^-] = 3.16 \times 10^{-6}$

$$[H_3O^+][OH^-] = 1 \times 10^{-14}$$

$$[H_3O^+] = 3.16 \times 10^{-9}$$

$$pH = -\log [H_3O^+]$$

$$pH = 8.5$$

6. a. Given: pH = 7

$$-\log [H_3O^+] = 7$$

$$\log [H_3O^+] = -7$$

$$[H_3O^+] = 1 \times 10^{-7}$$

$$[OH^-] = 1 \times 10^{-7}$$

b. Given: pH = 9.5

$$-\log [H_3O^+] = 9.5$$

$$[H_3O^+] = 3.16 \times 10^{-10}$$

$$[OH^-] = 3.16 \times 10^{-5}$$

c. $$[H_3O^+] = 2.0 \times 10^{-5}$$

$$[OH^-] = 5.0 \times 10^{-10}$$

d. Given: pOH = 3.6

$$-\log [OH^-] = 3.6$$

$$\log [OH^-] = -3.6$$

$$[OH^-] = 2.5 \times 10^{-4}$$

$$[H_3O^+] = 4.0 \times 10^{-11}$$

e. $$[OH^-] = 6.3 \times 10^{-10}$$

$$[H_3O^+] = 1.2 \times 10^{-5}$$

SECTION 11.7 Weak Acids

7. Write the equilibrium expression for the following
   weak acids:
   a. HF
   b. HCN
   c. H_2S
   d. HSO_3^-

8. A 0.10 molar solution of a weak acid is 2.00% ionized at equilibrium. What is the dissociation constant of the acid.

9. A 0.0100 molar solution of a weak acid is 5.00%

ionized at equilibrium. What is the dissociation constant of the acid?

10. What is the pH of a 0.1 M solution of HF ?
(K_a = 3.53 x 10^{-4})

11. What is the pH of a 0.05 M solution of an acid, HA , if the dissociation constant of the acid is 2.72 x 10^{-7} ?

SECTION 11.8 Weak Bases

12. Write the equilibrium expression for each of the following weak bases:
a. F^-
b. NH_3
c. SO_3^{2-}
d. $C_2H_3O_2^-$

13. What is the pH of a 0.1 M solution of Na_2SO_3 ?
(K_b for SO_3^{2-} is 1 x 10^{-7})

14. What is the pH of a 1.0 m solution of NaF ?
(K_b for F^- is 1.5 x 10^{-11})

Answers:

7. a. In order to write the equilibrium expression for a weak acid, we write the balanced equation for the dissociation of the acid in water:

$$HF + H_2O \rightleftharpoons H_3O^+ + F^-$$

We then write the mass action expression for the reaction by putting the concentration of the hydronium ion times the concentration of the anion of the acid (in this case F^-) in the numerator and the concentration of the undissociated acid in the denominator of a fraction. This fraction is called the mass action expression.

$$\frac{[H_3O^+][F^-]}{[HF]}$$

At equilibrium the mass action expression for the dissociation of the acid is equal to the dissociation constant of the weak acid.

$$K_a = \frac{[H_3O^+][F^-]}{[HF]}$$

Note: The concentration of water is omitted from the mass action expression because in dilute aqueous solutions the concentration of water is constant and is included in the dissociation constant of the weak acid.

b. $K_a = \dfrac{[H_3O^+][CN^-]}{[HCN]}$

c. $K_a = \dfrac{[H_3O^+][HS^-]}{[H_2S]}$

d. $K_a = \dfrac{[H_3O^+][SO_3^{2-}]}{[HSO_3^-]}$

8. Write a balanced equation for the dissociation of the weak acid:

$$HA + H_2O \rightleftharpoons H_3O^+ + A^-$$

Write the equilibrium expression for the weak acid:

$$K_a = \frac{[H_3O^+][A^-]}{[HA]}$$

Assign the equilibrium concentration of $[HA]$, $[H_3O^+]$ and $[A^-]$:
Original concentration of HA = 0.10 M.

If 2.00% of HA ionizes, the concentration of $[H_3O^+]$ and $[A^-]$ is

$$[H_3O^+] = [A^-] = 0.10 \text{ M} \times 0.02 = 0.002 \text{ M}$$

The equilibrium concentration of HA is

$$0.1 \text{ M} - 0.002 \text{ M} = 0.098 \text{ M}$$

Substitute the equilibrium concentrations into the equilibrium expression and solve it for the K_a :

$$K_a = \frac{(0.002)(0.002)}{0.098}$$

$$K_a = 4.1 \times 10^{-5}$$

9. The solution is the same as in Problem 8.

$$K_a = 2.6 \times 10^{-5}$$

10. Write a balanced equation for the dissociation of the acid:

$$HF + H_2O \rightleftharpoons H_3O^+ + F^-$$

Write the equilibrium expression for the dissociation:

$$K_a = \frac{[H_3O^+][F^-]}{[HF]}$$

Assign the equilibrium concentrations for each substance:

$$\text{Let} \quad x = [H_3O^+] = [F^-]$$

$$\text{Then} \quad HF = 0.10 - x$$

Substitute these values into the equilibrium expression for the dissociation and solve for x:

$$3.53 \times 10^{-4} = \frac{(x)(x)}{0.10 - x}$$

Since x is small, we will assume $0.10 - x = 0.10$

$$3.53 \times 10^{-4} = \frac{(x)(x)}{0.10}$$

$$x = 5.9 \times 10^{-3}$$

$$[H_3O^+] = 5.9 \times 10^{-3}$$

$$pH = 2.23$$

11. $[H_3O^+] = 1.2 \times 10^{-4}$

$$pH = 3.9$$

12. a. $F^- + H_2O \rightleftarrows HF + OH^-$

b. $NH_3 + H_2O \rightleftarrows NH_4{}^+ + OH^-$

c. $SO_3{}^{2-} + H_2O \rightleftarrows HSO_3^- + OH^-$

d. $C_2H_3O_2^- + H_2O \rightleftarrows H(C_2H_3O_2) + OH^-$

13. SO_3^{2-} is a weak base and will react with H_2O to produce OH^-.

Write the equation for the equilibrium between the weak base, SO_3^{2-}, and H_2O:

$$SO_3^{2-} + H_2O \rightleftarrows HSO_3^- + OH^-$$

Write the equilibrium expression for the reaction:

$$K_b = \frac{[HSO_3^-][OH^-]}{[SO_3^{2-}]}$$

Given: $K_b = 1 \times 10^{-7}$

Assign the values of the equilibrium concentrations of the substances:

$$[HSO_3^-] = [OH^-] = x$$

$$[SO_3^{2-}] = 0.10 = -x$$

Substitute the equilibrium concentration into the equilibrium expression and solve for the unknown concentration:

$$1 \times 10^{-7} = \frac{(x)(x)}{0.10 - x}$$

Since x is so small, we will assume that $0.10 - x = 0.10$.

$$1 \times 10^{-7} = \frac{(x)(x)}{0.10}$$

$$x = 1.0 \times 10^{-4}$$

$$[OH^-] = 1.0 \times 10^{-4}$$

$$pOH = 4.0 \qquad pH = 10.0$$

14. $[OH^-] = 3.9 \times 10^{-6}$

   pOH = 5.4

   pH = 8.6

SECTION 11.11

15. What is the normality of
    a. a 0.100 M solution of HCl
    b. a 0.0500 M solution of H_2SO_4
    c. a 0.2500 M solution of H_3PO_4 .

16. Exactly 27.92 ml of a 0.1725 M solution of NaOH was required to neutralize 25.00 ml of HCl solution. What is the concentration of the HCl solution?

17. Exactly 31.65 ml of a 0.2615 M solution of NaOH was required to neutralize 25.00 ml of HCl solution. What is the concentration of the HCl solution?

18. Exactly 1.849 g of oxalic acid, $H_2C_2O_4$, required 29.65 ml of NaOH to completely neutralize it. What is the concentration of the NaOH solution? (Oxalic acid produces 2 moles of hydronium ions for each mole of acid.)

19. Exactly 2.976 g of oxalic acid, $H_2C_2O_4$, required 24.72 ml of a NaOH solution to neutralize it. What is the concentration of the NaOH solution?

Answers:

15. a. HCl produces one mole of hydronium ions per mole of acid. The normality and the molarity are identical. Normality = 0.10 N .

    b. H_2SO_4 produces 2 moles of hydronium ions per mole of acid. The normality is twice the molarity. Normality = 0.1000 N .

    c. H_3PO_4 produces 3 moles of hydronium ions per mole of acid. The normality is three times the molarity. Normality = 0.7500 N .

16. moles of NaOH = 0.02792 liters x 0.1725 mole/liter

moles of HCl = 0.02500 liters x (x mole/liter)
moles of HCl = moles of NaOH
0.02792 liters NaOH x 0.1725 mole/liter
$$= 0.02500 \text{ liters x } (x \text{ mole/liter})$$
x mole/liter = 0.1926 mole/liter

17. HCl = 0.3311 M

18. $H_2C_2O_4 + 2 \text{ NaOH} \rightarrow Na_2C_2O_4 + 2 H_2O$

$$1.849 \text{ g } H_2C_2O_4 \text{ x } \frac{1 \text{ mole } H_2C_2O_4}{90.06 \text{ g } H_2C_2O_4} \text{ x } \frac{2 \text{ mole NaOH}}{1 \text{ mole } H_2C_2O_4}$$

= 0.04106 mole NaOH

$$\frac{0.04106 \text{ mole NaOH}}{0.02965 \text{ liters}} = 1.385 \text{ M NaOH}$$

19. 2.674 M NaOH

SECTION 11.13 Reactions of Ions in Solution

20. Write the molecular equation, the total ionic equation, and the net ionic equation for the reaction that occurs when the following solutions are mixed:
a. silver nitrate and sodium bromide
b. zinc nitrate and sodium sulfide
c. barium chloride and sodium phosphate
d. calcium chloride and sodium carbonate

SECTION 11.14   Solubility Products

21. Write the solubility product expression for each of the following salts.
a. $CaCO_3$
b. $Ca_3(PO_4)_2$
c. Bi_2S_3
d. $Fe(OH)_3$

22. What is the value of the solubility product of calcium sulfide if a saturated solution contains 0.210 g of CaS per liter of solution?

23. A saturated solution of strontium sulfate contains 0.113 g per liter. What is the value of the solubility product of $SrSO_4$?

24. Which substance is more soluble, iron(II) sulfide or manganese(II) sulfide? (See Table 11.6 in the text.)

25. What is the molar solubility of the following (see Table 11.6 of the text for K_{sp} values):
    a. $BaSO_4$
    b. CaF_2
    c. PbS
    d. ZnS

Answers:

20. a. silver bromide is insoluble

$$AgNO_3{}_{(aq)} + NaBr_{(aq)} \rightleftarrows AgBr_{(s)} + NaNo_3$$
(molecular equation)

$$Ag^+_{(aq)} + NO_3^-{}_{(aq)} + Na^+_{(aq)} + Br^-_{(aq)}$$
$$\rightleftarrows AgBr_{(s)} + Na^+_{(aq)} + NO_3^-{}_{(aq)}$$

$$Ag^+_{(aq)} + Br^-_{(aq)} \rightleftarrows AgBr_{(s)}$$

b. $Zn(NO_3)_2{}_{(aq)} + Na_2S_{(aq)} \rightleftarrows ZnS_{(s)} + 2\ NaNO_3{}_{(aq)}$

$Zn^{2+} + 2\ NO_3^- + 2\ Na^+ + S^{2-} \rightleftarrows ZnS_{(s)} + 2\ Na^+ + 2\ NO_3^-$

$Zn^{2+} + S^{2-} \rightleftarrows ZnS_{(s)}$

c. $3\ BaCl_2{}_{(aq)} + 2\ Na_3PO_4{}_{(aq)} \rightleftarrows Ba_3(PO_4)_2{}_{(s)}$
$$+ 6\ NaCl_{(aq)}$$

$3\ Ba^{2+} + 6\ Cl^- + 6\ Na^+ + 2\ PO_4^{3-} \rightleftarrows Ba_3(PO_4)_2$
$$+ 6\ Na^+ + 6\ Cl^-$$

$3\ Ba^{2+} + 2\ PO_4^{3-} \rightleftarrows Ba_3(PO_4)_2$

d. $CaCl_2{}_{(aq)} + Na_2CO_3{}_{(aq)} \rightleftarrows CaCO_3{}_{(s)} + 2\ NaCl_{(aq)}$

$Ca^{2+} + 2\ Cl^- + 2\ Na^+ + CO_3^{2-} \rightleftarrows CaCO_3{}_{(s)} + 2\ Na^+$
$$+ 2\ Cl^-$$

$$Ca^{2+} + CO_3^{2-} \rightleftharpoons CaCO_{3(s)}$$

21. a. Write an equation for the equilibrium between the solid and its ions in solution:

$$CaCO_3 \rightleftharpoons Ca^{2+} + CO_3^{2-}$$

From the balance equation write the solubility constant expression:

$$K_{sp} = [Ca^{2+}][CO_3^{2-}]$$

b. $Ca_3(PO_4)_{2(s)} \rightleftharpoons 3\ Ca^{2+} + 2\ PO_4^{3-}$

$$K_{sp} = [Ca^{2+}]^3[PO_4^{3-}]^2$$

c. $K_{sp} = [Bi^{3+}]^2[S^{2-}]^3$

d. $K_{sp} = [Fe^{3+}][OH^-]^3$

22. Write the equation for the equilibrium between the solid and its ions in solution:

$$CaS_{(s)} \rightleftharpoons Ca^{2+} + S^{2-}$$

Write the solubility product expression:

$$K_{sp} = [Ca^{2+}][S^{2-}]$$

Assign the molar concentration of the ions:

$$\frac{0.210\ g}{1\ liter} \times \frac{1\ mole\ CaS}{72.1\ g\ CaS} = 2.9 \times 10^{-3}\ M$$

$$[Ca^{2+}] = [S^{2-}] = 2.9 \times 10^{-3}$$

Substitute the values into the solubility product expression and solve:

$$K_{sp} = (2.9 \times 10^{-3})(2.9 \times 10^{-3})$$

$$K_{sp} = 8.5 \times 10^{-6}$$

23. $K_{sp} = 3.8 \times 10^{-7}$

24. The K_{sp} values for MnS and FeS are 7×10^{-16} and

4×10^{-19} , respectively. Both salts produce the same number of ions in solution. Therefore, the salt, MnS , that has the larger solubility product is more soluble.

25. a. Write the equation for the equilibrium reaction:

$$BaSO_{4(s)} \rightleftharpoons Ba^{2+} + SO_4^{2-}$$

Write the solubility product expression:

$$K_{sp} = [Ba^{2+}][SO_4^{2-}]$$

Assign equilibrium concentrations for the ions:

$$[Ba^{2+}] = [SO_4^{2-}] = x$$

Substitute the values into the solubility product expression and solve:

$$K_{sp} = 1.5 \times 10^{-9} \quad \text{(from Table 11.6)}$$

$$1.5 \times 10^{-9} = (x)(x)$$

$$x = 3.9 \times 10^{-5} = \text{molar solubility}$$

b. $CaF_{2(s)} \rightleftharpoons Ca^{2+} + 2 F^-$

$K_{sp} = [Ca^{2+}][F^-]^2$

$[Ca^{2+}] = x$

$[F^-] = 2x$

$4.0 \times 10^{-11} = (x)(2x)^2$

$4x^3 = 4.0 \times 10^{-11} = \text{molar solubility}$

$x = 2.2 \times 10^{-4}$

c. molar solubility = 8.4×10^{-15}

d. molar solubility = 1.6×10^{-11}

SELF-TEST
1. The conjugate base of water is _____.
2. The conjugate acid of ammonia, NH_3 , is _____.
3. If the hydronium ion concentration of an aqueous solution is 1 x 10^{-5} M , the hydroxide ion concentration is _____.
4. If the pH of a solution is 4.71, the hydronium ion concentration is _____.
5. The pH of a solution that has a hydronium concentration of 2.5 x 10^{-4} M is _____.
6. The dissociation constant of a weak acid that is 4.0% ionized in a 0.10 M solution is _____.
7. The pH of a 0.1 M solution of a weak acid with a dissociation constant of 1.5 x 10^{-6} is _____.
8. Hydrogen sulfide has a K_a of 9.1 x 10^{-8} and hydrogen cyanide has a K_a of 4.9 x 10^{-10} . Which is the stronger acid?
9. The pH of a 2.0 M solution of NH_3 (K_b = 1.8 x 10^{-5}) is _____.
10. The pH of a 5.0 x 10^{-4} solution of NaOH is _____.
11. Which of the following substances is least soluble? NiS (K_{sp} = 3 x 10^{-21}) or FeS (K_{sp} = 4 x 10^{-19})
12. To what exponential power is the concentration of the sulfide ion raised in the solubility expression for Al_2S_3 ?
13. What is the molar solubility of barium sulfate? (K_{sp} = 1.5 x 10^{-9})
14. What is the maximum concentration of the silver ion in a 0.10 M solution of AgBr ? (The K_{sp} of AgBr is 5.0 x 10^{-13} .)
15. A saturated solution of silver cyanide, AgCN , contains 0.00023 g of AgCN per liter of solution. What is the value of the K_{sp} for AgCN ?

Answers:

1. OH^-
2. NH_4^+
3. 1 x 10^{-9}
4. 2.0 x 10^{-5}
5. 3.60
6. 1.7 x 10^{-4}
7. 3.4
8. H_2S
9. 11.8
10. 10.7
11. NiS

12. 3
13. 3.9×10^{-5}
14. 5×10^{-12} M
15. 2.9×10^{-12}

chapter

12 Rates of Reactions and Chemical Equilibrium

OBJECTIVES
1. You should know what is meant by the term "rate of reaction" (Section 12.1).
2. You should be able to explain the collision theory of reactions (Section 12.2).
3. You should be able to draw an energy profile diagram for a given reaction and identify the energies of the reactants, products, and activated complex (Section 12.3).
4. You should know the meaning of activation energy and energy of reaction and be able to relate them to an energy profile diagram for a given reaction (Section 12.3).
5. You should be able to explain the effect of the nature of the reactants on the reaction rate (Section 12.4).
6. You should be able to explain the effect of the concentration of reactants on the reaction rate (Section 12.5).
7. You should be able to write the general form for a reaction rate equation (Section 12.5).
8. You should be able to write the experimental rate equation from experimental data for a given reaction (Section 12.5).
9. From a given experimental rate equation, you should be able to predict the effect of varying concentrations of the various reactants in a given reaction (Section 12.6).
10. You should be able to explain the effect of

temperature on reaction rates and relate the temperature effect to the collision theory of reactions
(Section 12.6).
11. You should be able to explain how a catalyst affects
the reaction rate (Section 12.7).
12. You should be able to explain the terms "reversible
reaction" and "dynamic equilibrium" (Section 12.8).
13. You should be able to write the mass action expression for any given reaction (Section 12.9).
14. You should be able to write the equilibrium constant
expression for any given reaction (Section 12.9).
15. You should be able to determine the equilibrium constant for any reaction from the equilibrium concentrations of the reactants and products (Section
12.10).
16. You should be able to determine the equilibrium concentration of a substance from the value of the
equilibrium constant and the concentrations of the
other reactants and products (Section 12.11).
17. You should be able to predict whether a given reaction will occur based on the value of the equilibrium constant and the concentrations of the reactants and product (Section 12.11).
18. You should know and be able to apply Le Chatelier's
Principle to equilibrium systems (Section 12.12).
19. You should be able to list the factors that affect
the equilibrium of a system and to explain what effect each has on the system of equilibrium (Section
12.13).

IMPORTANT
TERMS
AND
CONCEPTS

SECTION 12.1

Reaction rate   the rate at which the concentration of
the reactants or products change in a given unit of
time during a chemical reaction. Rates are usually
expressed in moles/liter-second.

Chemical kinetics   the study of the rates of reaction
and the mechanisms by which chemical reactions pro-
ceed.

SECTION 12.2

Collision theory of reactions   in order for a reaction
to occur:
a. there must be a collision between the reacting

molecules.
  b. the orientation of the colliding molecules must
     be such that a new bond can be formed.
  c. the collision between molecules must be energetic
     enough to break old bonds and form new bonds.

<u>Homogeneous system</u>   a system in which all of the mole-
  cules of the reactants and products are in the same
  phase.

<u>Heterogeneous system</u>   a system in which one reactant
  is in one phase and the other reactant or reactants
  are in a different phase.

Section 12.3

<u>Reaction intermediate</u> or <u>activated complex</u>   the com-
  bination of reactants in a bond making and bond
  breaking intermediate.

<u>Activation energy</u>   the energy required for the reac-
  tants to form the activated complex.

<u>Exoenergetic reaction</u>   a reaction that releases energy.

<u>Endoenergetic reaction</u>   a reaction that requires
  energy to take place.

Section 12.4

If the energy difference between the energy of the reactants and the energy of the activated complex is
  small, the activation energy is small and the reaction rate is fast. If the energy difference between
  the reactants and the activated complex is large,
  the activation energy is large and the reaction is
  slow.

Section 12.5

If the concentration of the reactants that form the
  activated complex is increased, the rate of reaction
  increases.

<u>Rate equation</u>   an experimentally determined equation
  that shows the concentrations of which substances
  the rate is dependent on.

<u>Specific rate constant</u>   the proportionality constant
  between the rate of reaction and the concentration
  of the substances on which it is dependent.

Section 12.6

An increase in temperature always increases the reaction rate. According to the kinetic molecular theory, an increase in temperature increases the rate of molecular collisions and also increases the number of these collisions that have enough kinetic energy to cause bond making and bond breaking.

Section 12.7

Catalyst   a substance that increases the rate of reaction without undergoing any permanent change in the course of a reaction. The catalyst increases the rate of reaction by modifying the activated complex and lowering the activation energy of the reaction.

Enzymes   biological catalysts.

Section 12.8

Dynamic equilibrium   the seemingly static situation that exists when two opposing reactions are taking place at exactly the same rate.

Section 12.9

Mass action expression   the mathematical product of the concentrations of the substances on the right of a chemical equation raised to the exponential power of their coefficients divided by the mathematical product of the concentrations of the substances on the left of the equation raised to the exponential power of their coefficients.   For the equation

$$4\ A + 3\ B \rightleftarrows 2\ C + 3\ D$$

the mass action expression is

$$\frac{[C]^2[D]^3}{[A]^4[B]^3}$$

Equilibrium constant   the numerical value of the mass action expression for a system at equilibrium.

Section 12.10

See the Problems for this section.

Section 12.11

See the Problems for this section.

Section 12.12

Le Chatelier's Principle   when a system is at equilibrium and a condition that affects the equilibrium is changed, the system will always react in such a way as to tend to counteract the change.

Section 12.13

The factors affecting equilibrium are:
   a. the concentrations of the reactants and the products.
   b. temperature changes for exothermic and endothermic reactions.

Endothermic reaction   a reaction which absorbs heat.

Exothermic reaction   a reaction which releases heat.

QUESTIONS AND PROBLEMS

Section 12.1   Rates of Reactions

1. Draw an energy profile diagram for an endoenergetic reaction and indicate:
   a. the energy content of the reactants;
   b. the energy content of the activated complex;
   c. the energy content of the products;
   d. the activation energy;
   e. the energy of the reaction.

2. For the reaction:

$$H_2O_2 + 3 \ I^- + 2 \ H \rightleftarrows 2 \ H_2O + I_3^-$$

the experimental rate law is:

$$rate = k[H_2O_2][I^-]$$

What effect would the following changes have on the

initial reaction rate?
a. Changing the concentration of H^+.
b. Doubling the concentration of H_2O_2.
c. Decreasing the concentration of I^- by one half.
d. Doubling the concentration of H_2O_2 and HI.
e. Increasing the temperature of the reaction.

3. A kinetics study of the reaction:

$$2\ NO_{(g)} + 2\ H_{2(g)} \rightleftarrows N_{2(g)} + 2\ H_2O_{(g)}$$

yielded the following data at 1000 K:

Experiment	$[H_2]$ x 10^{-3}	$[NO]$ x 10^{-3}	Initial Rate 10^{-6} M/sec
1	6	2	4
2	6	4	16
3	3	4	8

What is the experimental rate law?

Answers:

1.

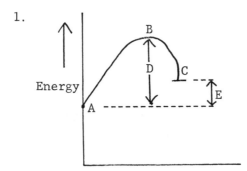

A, energy of reactants
B, energy of activated complex
C, energy of products
D, activation energy
E, energy of reaction

2. a. Changing the concentration of H^+ would have no
       effect on the reaction rate since it is not

in the rate equation.
  b. Doubling the concentration of H_2O_2 would double the rate.
  c. Halving the concentration of I^- would halve the reaction rate.
  d. Doubling the concentration of both H_2O_2 and I^- would make the reaction four times as fast, $(2)(2) = 4$.
  e. Increasing the temperature always increases the rate of a reaction.

3. Comparing the differences in the concentration of the reactants in the first two experiments, we can see that doubling the concentration of NO increases the rate 4 times. Therefore, the rate is dependent on the square of the concentration of NO.

Comparing experiments 2 and 3, we can see that doubling the concentration of H_2 doubles the rate. Therefore, the rate is dependent on the concentration of H_2. Putting these values into the experimental rate equation gives:

$$rate = k \, [NO]^2 \, [H_2]$$

SECTION 12.9   Chemical Equilibrium

4. Write the equilibrium expressions for the following reactions:

  a. $2 \, SO_{2(g)} + O_{2(g)} \rightleftharpoons 2 \, SO_{3(g)}$

  b. $2 \, O_{3(g)} \rightleftharpoons 3 \, O_{2(g)}$

  c. $CO_{(g)} + H_2O_{(g)} \rightleftharpoons CO_{2(g)} + H_{2(g)}$

  d. $CaSO_{3(s)} \rightleftharpoons CaO_{(s)} + SO_{2(g)}$

5. When 5.0 moles of HI was placed in a 1 liter reactor at $450°C$, 24% of the HI decomposed to form I_2 and H_2 at equilibrium. What is the equilibrium constant for the reaction?

6. If the equilibrium constant for the reaction:

$$H_{2(g)} + I_{2(g)} \rightleftharpoons 2 \, HI_{(g)}$$

is 50 at $450^{\circ}C$, what is the equilibrium concentration of each of the gases if 2 moles of H_2 and 2
moles of I_2 are placed in a 1 liter reactor and
heated to $450^{\circ}C$?

7. In the reaction:

$$2\ O_{3(g)} \rightleftarrows 3\ O_{2(g)} + Heat$$

if the system is at equilibrium, what effect will
the following changes have on the equilibrium?
a. The pressure of the bases is increased by decreasing the volume of the reactor.
b. The concentration of O_2 is reduced.
c. The temperature of the reaction is increased.
d. A catalyst is added.

Answers:

4. a. All of the substances are in the gas phase so all
      of them will be used in the mass action expres-
      sion.   The numerator of the mass action expres-
      sion is the concentration of SO_3 raised to the
      second power.   The denominator of the mass action
      expression is the concentration of SO_2 raised to
      the second power times the concentration of O_2 :

$$K_{eq} = \frac{[SO_3]^2}{[SO_2]^2\ [O_2]}$$

   b.
$$K_{eq} = \frac{[O_2]^3}{[O_3]^2}$$

   c.
$$K_{eq} = \frac{[CO_2][H_2]}{[CO][H_2O]}$$

   d.
$$K_{eq} = [SO_2]$$

Note: Since $CaSO_3$ and CaO are both solids, their
respective concentrations are constant and included
in the value of the equilibrium constant rather than
in the mass action expression.

5. Write an equation for the reaction:

$$2\ HI \rightleftarrows H_2 + I_2$$

Assign the equilibrium concentrations: if 24% of the HI decomposes, 1.2 moles of HI decompose and 0.6 moles of H_2 and I_2 are formed.

At equilibrium, [HI] = 3.8 mole/liter; $[H_2] = [I_2]$ = 0.6 mole/liter.

Write the equilibrium expression for the reaction:

$$K_{eq} = \frac{[I_2][H_2]}{[HI]^2}$$

Substitute the equilibrium concentrations into the mass action expression and solve the equation:

$$K_{eq} = \frac{(0.6)(0.6)}{(3.8)^2}$$

$$K_{eq} = 0.025$$

6. Write the equilibrium expression for the reaction:

$$K_{eq} = \frac{[HI]^2}{[H_2][I_2]}$$

Assign the equilibrium concentrations to the reactants and products: We will assume that x moles of H_2 and I_2 are converted to $2x$ moles of HI

$[H_2] = 2 - x$ mole/liter $[I_2] = 2 - x$ mole/liter
$[HI] = 2x$ mole/liter $K_{eq} = 50$

Substituting into the equilibrium expression:

$$50 = \frac{(2 - x)(2 - x)}{(2x)^2}$$

This is a quadratic equation and to simplify it we will take the square root of each side:

$$50 = \frac{(2 - x)(2 - x)}{(2x)^2}$$

$$7.1 = \frac{2 - x}{2x} \qquad x = 0.13 \text{ mole/liter}$$

$[H_2] = [I_2] = 1.87$ mole/liter;   [HI] = 0.13 mole/liter.

7. a. The reaction will go to the left to occupy less
      volume according to Le Chatelier's Principle.
   b. The reaction will shift to the right to replace
      the O_2 removed, according to Le Chatelier's Prin-
      ciple.
   c. The reaction will shift to the left in order to
      absorb the heat.
   d. A catalyst has no effect on the position of equi-
      librium.

SELF-TEST 1. List two methods of increasing the rate of a chemi-
              cal reaction.

           Given the reaction:

$$2 \, A + 2 \, B \rightleftharpoons C + 4 \, D$$

           with the experimental rate equation:

$$\text{rate} = k \, [A][B]^2$$

2. Doubling the concentration of A will do what to the
   reaction rate?
3. Doubling the concentration of B will do what to the
   reaction rate?

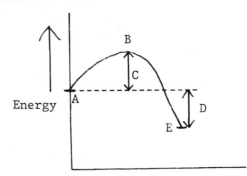

Given the energy profile diagram shown above, identify:
4. A
5. B
6. C
7. D
8. E
9. Is the reaction exoenergetic or endoenergetic?

10. Write the equilibrium expression for the reaction:

$$NH_{3(g)} + HCl_{(g)} \rightleftarrows NH_4Cl_{(s)}$$

11. Write the equilibrium expression for the reaction:

$$2\ Cl_{2(g)} + 2\ H_2O_{(g)} \rightleftarrows 4\ HCl_{(g)} + O_{2(g)}$$

12. At $1000^\circ C$ the equilibrium constant for the reaction:

$$H_{2(g)} + CO_{2(g)} \rightleftarrows H_2O_{(g)} + CO_{(g)}$$

is 1.60. What is the equilibrium concentration of
each gas if the initial concentration of H_2 and CO_2
are each 2 M?

13. For the reaction:

$$H_{2(g)} + I_{2(g)} \rightleftarrows 2\ HI_{(g)} \qquad 450^\circ C$$

the equilibrium concentrations are:
$H_2 = I_2 = 0.50$ M;   HI = 3.75 M.

What is the equilibrium constant for the reaction?

14. For the reaction,

$$Heat + N_2O_{4(g)} \rightleftarrows 2\ NO_{2(g)}$$

at equilibrium what effect will the following
changes have on the concentration of N_2O_4?
a. a catalyst is added to the system;
b. the pressure is increased by decreasing the size
   of the reactor;
c. more NO_2 is added to the system;
d. the temperature of the reaction is decreased.

Answers:

1. Increasing the temperature and adding a catalyst
   will always increase the rate of reaction. Increasing the concentration of the reactants usually increases the rate, but the increase is dependent on
   the nature of the reaction.
2. Double the rate
3. Increase the rate 4 times

4. Energy of reactants
5. Energy of the activated complex
6. Activation energy
7. Energy of reaction
8. Energy of products
9. The reaction is exoenergetic

10. $K_{eq} = \dfrac{1}{[NH_3][HCl]}$

11. $K_{eq} = \dfrac{[HCl]^4 \, [O_2]}{[Cl_2]^2 \, [H_2O]^2}$

12. $[H_2] = [CO_2] = 0.77 \; M \,;$ $[H_2O] = [CO] = 1.23 \; M$

13. $K_{eq} = 56$

14. a. None.
    b. $N_2O_{4(g)}$ concentration increases.
    c. $N_2O_{4(g)}$ concentration increases.
    d. $N_2O_{4(g)}$ concentration increases.

chapter

13 Oxidation–Reduction Reactions

OBJECTIVES
1. You should be able to explain what is meant by oxidation and reduction (Section 13.1).
2. In a given equation you should be able to determine: the substance oxidized, the substance reduced, the oxidizing agent, and the reducing agent (Section 13.1).
3. You should know the rules for assigning oxidation numbers (Section 13.2).
4. You should be able to assign oxidation numbers to any atom in any compound (Section 13.2).
5. You should be able to balance oxidation - reduction equations by the half reaction method and by the oxidation number method (Section 13.3).
6. You should be able to explain the operation of a voltaic cell (Section 13.4).
7. You should be able to explain the operation of an electrolytic cell (Section 13.5).
8. You should be able to calculate the voltage of a standard cell from a table of standard reduction potentials (Section 13.6).

IMPORTANT TERMS AND CONCEPTS

SECTION 13.1

Oxidation   the loss of electrons by an atom.

Reduction   the gain of electrons by an atom.

Oxidizing agent   the substance that takes electrons during the oxidation process.

Reducing <u>agent</u>   the substance that supplies electrons during the reduction process.

SECTION 13.2

See Questions and Problems for this section.

SECTION 13.3

See Questions and Problems for this section.

SECTION 13.4

<u>Voltaic</u> <u>cell</u>   a device for the conversion of chemical energy to electrical energy.

<u>Anode</u>   the electrode at which oxidation takes place.

<u>Cathode</u>   the electrode at which reduction takes place.

SECTION 13.5

<u>Electrolytic</u> <u>cell</u>   a device for the conversion of electrical energy to chemical energy.

SECTION 13.6

<u>Standard</u> <u>reduction</u> <u>potential</u>   the voltage required to reduce a substance compared to the reduction of H^+ to H_2 when all of the substances are in their standard state.

QUESTIONS
AND
PROBLEMS

1. Assign oxidation numbers to each atom in the following compounds or ions:

a. CBr_4

b. $H_2C_2O_4$

c. NO_3^-

d. N_2O_4

e. O_2F_2

f. IF_5

g. $Na_2S_2O_8$

h. Hg_2Cl_2

i. Fe_2O_3

j. $K_2Cr_2O_7$

2. Balance the following equations:

   a. $Cu + HNO_3 \rightarrow Cu(NO_3)_2 + NO + H_2O$

   b. $H_2S + HNO_3 \rightarrow S + NO + H_2O$

   c. $PbO_2 + Sb + NaOH \rightarrow PbO + NaSbO_2 + H_2O$

   d. $MnO_4^- + S_2O_3^{2-} \rightarrow S_4O_6^{2-} + Mn^{2+}$ (acid solution)

   e. $H_4CO + MnO_4^- \rightarrow CO_3^{2-} + Mn^{2+}$ (basic solution)

3. Which of the following reactions will occur in the direction they are written: (Use Table 13.2 to make your predictions.)

   a. $Cu + Zn^{2+} \rightarrow Cu^{2+} + Zn$

   b. $Fe^{2+} + Zn \rightarrow Fe + Zn^{2+}$

   c. $Sn^{2+} \rightarrow Sn + Sn^{4+}$

   d. $Ag^+ + Cl^- \rightarrow Ag + Cl_2$

   e. $Cl^- + MnO_4^- + H^+ \rightarrow Cl_2 + Mn^{2+} + H_2O$

Answers:

1. a. $C = 4+$; $Br = 1-$
   b. $H = 1+$; $C = 2+$; $O = 2-$
   c. $N = 5+$; $O = 2-$
   d. $N = 2+$; $O = 2-$
   e. $O = 1+$; $F = 1-$
   f. $I = 5+$; $F = 1-$
   g. $Na = 1+$; $S = 7+$; $O = 2-$
   h. $Hg = 1+$; $Cl = 1-$
   i. $Fe = 3+$; $O = 2-$
   j. $K = 1+$; $Cr = 6+$; $O = 2-$

2. a. Cu goes from a 0 oxidation state to a 2+ oxidation state.   N goes from a 5+ oxidation state in HNO_3 to a 2+ oxidation state in NO .   Cu loses $2e^-$;   N gains $3e^-$:

$$\overset{\text{3 x loss 2e}^-}{3 \text{ Cu} + 2 \text{ HNO}_3 \rightarrow 3 \text{ Cu(NO}_3)_2 + 2 \text{ NO} + \text{H}_2\text{O}}$$

2 x gain 3e⁻

6 extra HNO_3 molecules must be added to the left side to balance the 6 NO_3 that are with Cu in $Cu(NO_3)_2$.

$$3 \text{ Cu} + 8 \text{ HNO}_3 \rightarrow 3 \text{ Cu(NO}_3)_2 + 2 \text{ NO} + \text{H}_2\text{O}$$

The hydrogens are balanced by adding 4 H_2O molecules to the right side of the equation:

$$3 \text{ Cu} + 8 \text{ HNO}_3 \rightarrow 3 \text{ Cu(NO}_3)_2 + 2 \text{ NO} + 4 \text{ H}_2\text{O}$$

A check of the oxygens shows 16 on each side. The equation is balanced.

b.

2 x gain 3e⁻

$$\text{H}_2\text{S} + \text{HNO}_3 \rightarrow \text{S} + \text{NO} + \text{H}_2\text{O}$$

3 x loss 2e⁻

S goes from a 2- to a 0; N goes from 5+ to 2+.

$$3 \text{ H}_2\text{S} + 2 \text{ HNO}_3 \rightarrow 3 \text{ S} + 2 \text{ NO} + 2 \text{ H}_2\text{O}$$

Balance the H atoms by increasing the number of H_2O molecules to 4 H_2O:

$$3 \text{ H}_2\text{S} + 2 \text{ HNO}_3 \rightarrow 3 \text{ S} + 2 \text{ NO} + \text{H}_2\text{O}$$

A check of the oxygens shows 6 on each side; the equation is balanced.

c. $3 \text{ PbO}_2 + 2 \text{ Sb} + 2 \text{ NaOH} \rightarrow 3 \text{ PbO} + 2 \text{ NaSbO}_2 + \text{H}_2\text{O}$

d. Mn goes from a 7+ in MnO_4^- to a 2+ in Mn^{2+} which is a gain of 5e⁻. S goes from a 2+ in $S_2O_3^{2-}$ to 2.5+ in $S_4O_6^{2-}$ which is the loss of $(1/2)e^-$ for each sulfur.

20 x loss 1/2 e⁻

$$2 \text{ MnO}_4^- + 10 \text{ S}_2\text{O}_3^{2-} \rightarrow 5 \text{ S}_4\text{O}_6^{2-} + 2 \text{ Mn}^{2+}$$

2 x gain 5e⁻

In acid solution balance the charge with H^+.

$$16 \ H^+ + 2 \ MnO_4^- + 10 \ S_2O_3^{2-} \rightarrow 5 \ S_4O_6^{2-} + 2 \ Mn^{2+}$$

Balance the H atoms with H_2O:

$$16 \ H^+ + 2 \ MnO_4^- + 10 \ S_2O_3^{2-} \rightarrow 5 \ S_4O_6^{2-}$$
$$+ 2 \ Mn^{2+} + 8 \ H_2O$$

A check of the oxygens shows 38 on each side; the equation is balanced.

e. C goes from a 2- oxidation state to a 4+ state for a loss of $6e^-$. Mn goes from a 7+ to a 2+ for a gain of $5e^-$.

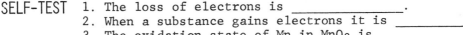

$$5 \ H_4CO + 6 \ MnO_4^- \rightarrow 5 \ CO_3^{2-} + 6 \ Mn^{2+}$$

In basic solution the charge is balanced by adding OH^-.

$$5 \ H_4CO + 6 \ MnO_4^- \rightarrow 5 \ CO_3^{2-} + 6 \ Mn^{2+} + 8 \ OH^-$$

The H atoms are balanced by adding H_2O.

$$5 \ H_4CO + 6 \ MnO_4^- \rightarrow 5 \ CO_3^{2-} + 6 \ Mn^{2+} + 8 \ OH^-$$
$$+ 6 \ H_2O$$

A check of the oxygens shows 29 on each side; the equation is balanced.

3. a. The reverse reaction will take place.
   b. The reverse reaction will take place.
   c. The reverse reaction will take place.
   d. The reverse reaction will take place.
   e. The reaction will go as written.

SELF-TEST   1. The loss of electrons is _____.
             2. When a substance gains electrons it is _____.
             3. The oxidation state of Mn in MnO_2 is _____.

4. The oxidation state of C in H_2CF_2 is _____.
5. The oxidation state of P in H_3PO_2 is _____.

In the equation: $Cl^- + MnO_4^- + H^+ \rightarrow Cl_2 + Mn^{2+}$:

6. The coefficient for Cl^- in the balanced equation is _____.

7. The coefficient for Cl_2 in the balanced equation is _____.

8. The coefficient for H^+ in the balanced equation is _____.

After the equation

$$HClO_2 + Cl^- \rightarrow HClO \qquad (acid\ solution)$$

is balanced:

9. The coefficient for H^+ is _____.
10. The coefficient for HClO is _____.

Use Table 13.2 in the text for Questions 11 and 12.

11. Will H_2 gas be evolved when an iron nail is dropped into a 1 M solution of HCl ?
12. Is silver iodide, AgI , stable?

Answers:

1. Oxidation
2. Reduced
3. 4+
4. 0
5. 1+
6. 10
7. 5
8. 16
9. 1
10. 2
11. Yes
12. No